FEEDBACK CONTROL IN SYSTEMS BIOLOGY

Carlo Cosentino | Declan Bates

CRC Press
Taylor & Francis Group
Boca Raton London New York

CRC Press is an imprint of the
Taylor & Francis Group, an **informa** business

A CHAPMAN & HALL BOOK

CRC Press
Taylor & Francis Group
6000 Broken Sound Parkway NW, Suite 300
Boca Raton, FL 33487-2742

First issued in paperback 2019

ISBN-13: 978-1-4398-1690-5 (hbk)
ISBN-13: 978-0-367-38226-1 (pbk)

Library of Congress Cataloging-in-Publication Data

Cosentino, Carlo.
    Feedback control in systems biology / Carlo Cosentino, Declan Bates.
        p. ; cm.
    Includes bibliographical references and index.
    ISBN 978-1-4398-1690-5 (hardcover : alk. paper)
    I. Bates, Declan. II. Title.

[DNLM: 1. Systems Biology. 2. Feedback, Physiological. 3. Models, Biological. QU 26.5]
611'.0181--dc23   2011036516

To my parents, Rosa and Nicola, and to Cinzia.

– C.C.

To my parents, Bríd and Tony, and to Orlando and Lauren.

– D.B.

Contents

Preface

The field of systems biology encompasses scientists with extremely diverse backgrounds, from biologists, biochemists, clinicians and physiologists to mathematicians, physicists, computer scientists and engineers. Although many of these researchers have recently become interested in control-theoretic ideas such as feedback, stability, noise and disturbance attenuation, and robustness, it is still unfortunately the case that only researchers with an engineering background will usually have received any formal training in control theory. Indeed, our initial motivation to write this book arose from the difficulty we found in recommending an introductory text on feedback control to colleagues who were not from an engineering background, but who needed to understand control engineering methods to analyse complex biological systems.

This difficulty stems from the fact that the traditional audience for control textbooks is made up of electrical, mechanical, process and aerospace engineers who require formal training in control system *design methods* for their respective applications. Systems biologists, on the other hand, are more interested in the fundamental concepts and ideas which may be used to analyse the effects of feedback in evolved biological control systems. Researchers with a biological sciences background may often also lack the expertise in physical systems modelling (Newtonian mechanics, Kirchhoff's electrical circuit laws, etc.) that is typically assumed in the examples used in standard texts on feedback control theory. The type of "control applications" in which a systems biologist is interested are systems such as metabolic and gene-regulatory networks, not aircraft, robots or engines, and the type of mathematical models they are familiar with are typically derived from classical reaction kinetics, not classical mechanics.

Another significant problem for systems biologists is that current undergraduate books on control theory (which introduce the basic concepts at great length) are uniformly restricted to linear systems, while nonlinear systems are usually only considered by specialist postgraduate texts which require advanced mathematical skills. Although it will always be appropriate to introduce basic ideas in control using linear systems, biological systems are in general highly nonlinear, and thus a clear understanding of the effects of nonlinearity is crucial for systems biologists.

To address these issues, we have tried to write a text on feedback control for systems biologists which

- is self-contained, in that it assumes no prior exposure to systems and control theory;

- focuses on the essential ideas and concepts from control theory that have found applicability in the systems biology research literature, including basic linear introductory material but also more advanced nonlinear techniques;

- uses examples from cellular and molecular biology throughout to illustrate key ideas and concepts from control theory;

- is concise enough to be used for self-study or as a recommended text for a single advanced undergraduate or postgraduate module on feedback control in a course on biological science, bioinformatics, systems biology or bioengineering.

During the time we have spent preparing this book we have also been struck by the constantly increasing interest among control engineers in biological systems, and thus a second goal of this text has been to provide an overview of how the many powerful tools and techniques of control theory may be applied to analyse biological networks and systems. Although we do assume that the reader has some familiarity with basic modelling concepts for biological systems, such as mass-action and Michaelis–Menten kinetics, we have provided introductory descriptions of many of the biological systems considered in the book, in the hope of enticing many more control engineering researchers into systems biology.

The book is made up of eight chapters. Chapter 1 provides an introduction to some basic concepts from feedback control, discusses some examples of biological feedback control systems and gives a brief historical overview of previous attempts to apply feedback control theory to analyse biological systems. Chapters 2 and 3 introduce a number of fundamental tools and techniques for the analysis of linear and nonlinear systems, respectively. Fundamental concepts such as state-space models, frequency domain analysis, stability and performance are introduced in the context of linear systems, while Chapter 3 discusses more advanced notions of stability for nonlinear systems, and also provides an overview of numerical optimisation methods for the analysis of complex nonlinear models. Chapter 4 focusses on the role of negative feedback in biological processes, and introduces notions of robustness, integral control and performance tradeoffs. Chapter 5 considers the rich variety of dynamics which arise due to positive feedback, and introduces tools such as bifurcation diagrams, monotone systems theory and chemical reaction network theory, which can be used to analyse bistable and oscillatory systems. Chapter 6 focusses on the issue of robustness, and provides an overview of the available robustness analysis methods, such as sensitivity analysis, μ-analysis, sum-of-squares polynomials and Monte Carlo simulation, which may be used to assist in validating or invalidating models of biological systems. A range of techniques for the reverse-engineering of biological interaction networks is described in Chapter 7. These techniques, which are rooted in the branch of control engineering known as system identification, appear to have huge potential to complement and augment the statistical approaches for network inference which currently dominate research on biological interaction networks. Finally, Chapter 8 provides an introduction to the analysis of stochastic biological control systems, and points out some exciting new research directions for control theory which are directly motivated by the particular dynamic characteristics of biological systems.

A key feature of the book is the use of biological case studies at the end of each chapter, in order to provide detailed examples of how the techniques introduced in the previous sections may be applied to analyse realistic biological systems. Each case study starts with an introductory section which provides a simple explanation of the relevant biological background, so that readers from the physical sciences can quickly understand the key features of the chosen system.

By its very nature, this book cannot pretend to provide an exhaustive treatment of all aspects of control theory. It does, however, represent a first attempt to arrange in a pedagogical manner the methods and applications of control theory that pertain to systems biology. Our aim has been to pinpoint the most important achievements to date, provide a useful reference for current researchers, and present a sound starting point for young scientists entering this exciting new field.

Acknowledgements

This book would not have been written without the help of many people, whom we would like to thank. First of all, we thank our better halves, Cinzia and Lauren, for their constant support and for having tolerated us spending holidays and nights dedicated to this book over the last two years. Our gratitude goes also to our mentors: C.C. would like to thank Francesco Amato for having instilled his genuine passion for research in control theory and for being both a stimulating and friendly guide; D.B. thanks Ian Postlethwaite for the same reasons.

Most of all we would like to acknowledge the numerous contributions of our current and former students and colleagues to the content and format of this book. In particular, we would like to thank Francesco Montefusco (formerly University of Catanzaro, now University of Exeter), Alessio Merola, Antonio Passanti, Luca Salerno and Basilio Vescio (University of Catanzaro), Jongrae Kim (University of Glasgow), Najl Valeyev (University of Kent), Pat Heslop-Harrison and Nick Brindle (University of Leicester), William Bryant (Imperial College London), Ian Stansfield, Claudia Rato and Heather Wallace (University of Aberdeen), Jon Hardman (University of Nottingham), Kwang-Hyun Cho (KAIST), and Prathyush Menon, Anup Das, Svetlana Amirova,

Orkun Soyer and Ozgur Akman (University of Exeter).

We thank all the colleagues who have provided us with feedback, comments, corrections or just useful discussions, which have helped, either directly or indirectly, to greatly improve this book.

Carlo Cosentino and Declan Bates
Catanzaro and Exeter

"The footsteps of Nature are to be traced, not only in her ordinary course, but when she seems to be put to her shifts, to make many doublings and turnings, and to use some kind of art in endeavouring to avoid our discovery."

- Robert Hooke, Micrographia, 1665.

"Nessuno effetto è in natura sanza ragione; intendi la ragione e non ti bisogna sperienzia."

- Leonardo da Vinci (1452 1519).

1

Introduction

In this chapter, we introduce some general concepts from control engineering, and describe a number of biological systems in which feedback control (often referred to as "regulation" in the biological literature) plays a fundamental role. We discuss the history of applying control theoretic techniques to the analysis of biological systems, and highlight the recent renewed interest in this area as a result of the explosive growth in the modelling of biological processes in current systems biology research.

1.1   What is feedback control?

A control system may be defined as an interconnection of components forming a configuration that provides a desired response. In an open-loop control system, as shown in Fig. 1.1, a controller C sends a signal to an actuating device (or actuator) A which can modify the state of a process P to obtain the desired output response. In the case of engineering systems, the process to be controlled (often called a *plant* in control engineering terminology) is generally taken as fixed, while the controller represents that part of the system which is to be designed by the engineer. For example, in the design of an aircraft flight control system, the plant P would represent the dynamics of the aircraft, the actuator would correspond to the aerodynamic control surfaces (rudders, ailerons and flaps) and the controller would be the computerised autopilot whose function is to maintain a steady flight trajectory.

If the dynamics of the plant are perfectly known (i.e. there existed a "perfect" model of its dynamics), and the control system is not subject to any environmental disturbances, then the output of the plant y could in theory be made to perfectly track any desired reference signal r using open-loop control simply by setting $C = (PA)^{-1}$ so that $y = (PA)Cr = (PA)(PA)^{-1}r = r$. In practice, however, neither of the above conditions ever holds, since even the most advanced model of the dynamics of the plant and actuators will always deviate to some extent from their actual behaviour (which may also change over time) and almost all real control systems are subject to significant disturbance inputs from their environments which alter their dynamic behaviour. As shown in Fig. 1.2, the effect of such plant uncertainty Δ and disturbances

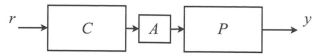

FIGURE 1.1: Open-loop control system

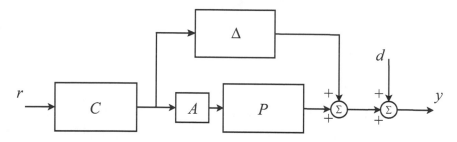

FIGURE 1.2: Open-loop control system with plant uncertainty and disturbances.

d is to make $y = d + (PA + \Delta)Cr$. Since both Δ and d are unknown, it is thus not possible to design an open-loop controller C to make $y = r$. It is the inevitable presence of *uncertainty* in both the dynamics of the process to be controlled, and the environment in which it operates, that necessitates the use of feedback control.

As shown in Fig. 1.3, a closed-loop feedback control system uses a sensor S to continuously "feed back" a measurement of the actual output of the system. This signal is then compared with the desired output to generate an error signal — this error signal forms the input to the controller which in turn generates a control signal which is input to the plant. Large error signals result in large control inputs which tend to bring the output of the plant closer to its desired state, in turn reducing the error. Now consider the effects of plant uncertainty and disturbances on this system, as shown in Fig. 1.4. The output of the closed-loop system is given by:

$$y = d + (\Delta + PA)C(r - y)$$
$$= d + (\Delta + PA)Cr - (\Delta + PA)Cy$$
$$\Rightarrow y[1 + (\Delta + PA)C] = d + (\Delta + PA)Cr$$
$$\Rightarrow y = \frac{1}{1 + (\Delta + PA)C}\, d + \frac{(\Delta + PA)C}{1 + (\Delta + PA)C}\, r$$

Notice that the controller C is now able to directly attenuate the effects of uncertainty on y — indeed, as $C \to \infty$ then $y \to r$ for *any* finite values of d and Δ. Of course, our analysis here is extremely simplistic since we are neglecting transient dynamics, the fact that measurement sensors are not

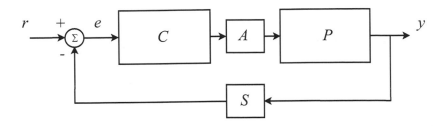

FIGURE 1.3: Closed-loop feedback control system.

perfect (i.e. $S \neq 1$) as well as a host of other limitations imposed by the particular dynamical properties of the plant and controller. The fundamental point remains, however, that it is the power of feedback to combat uncertainty (ensure *robustness*) which makes it so useful for the purposes of control.

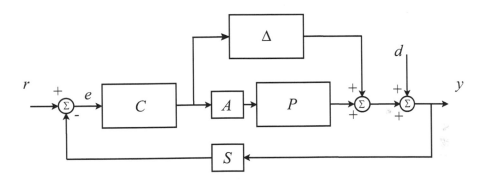

FIGURE 1.4: Closed-loop feedback control system with plant uncertainty and disturbances.

The vast majority of engineered control systems are negative feedback systems of the type shown in Fig. 1.4, i.e. the feedback signal is subtracted from the reference signal to generate the error signal for the controller. These types of control systems are generally employed to maintain systems at a particular set-point or to track dynamic reference signals. Many biological systems have evolved to also exploit *positive feedback*, for the purposes of signal amplification, noise suppression and to generate complex dynamics such as bistability and oscillations — these applications of feedback will be considered in detail in Chapter 5.

1.2   Feedback control in biological systems

In common with engineering systems, biological systems are also required to operate effectively in the presence of internal and external uncertainty (e.g. genetic mutations and temperature changes, respectively). It is thus not surprising that evolution has resulted in the widespread use of feedback, and research over the last decade in systems biology has highlighted the ubiquity of feedback control systems in biology, [1]-[7]. Due to the scale and complexity of biology, the resulting control systems often take the form of large interconnected regulatory networks, in which it is sometimes difficult to distinguish the "process" that is being controlled from its regulatory component or "controller."

Two remarks are in order here. First, it should be appreciated that, although feedback control theory typically assumes the existence of a separate plant and controller, most theoretical results and analysis tools do not require such a separation and can be formulated in terms of the open-loop or closed-loop transfer functions of the system. Second, there can often be significant advantages in attempting to conceptually separate the different components of a biological control system into functional modules with clearly defined roles, as this allows subsequent analysis of the system's dynamics to more clearly identify the role of the network structure in delivering the required system-level performance, [8].

Consider, for example, the different effects on phenotypic responses that may arise due to variations in the structure of a biological network versus changes in its "parameters" from their normal physiological values. An example of structural perturbation is the elimination of autoregulatory loops during transcription, which has been demonstrated to cause an increased variance in the *in vivo* protein expression resulting in phenotypic variability, [3]. On the other hand, a mutation in one copy of the NF1 gene constitutes an example of a parametric perturbation which results in higher incidences of benign tumours due to an increased noise-to-signal ratio caused by haploinsufficiency, [9, 10]. Understanding the structural design principles of such biological systems will be crucial to the development of effective therapeutic strategies to counteract disease, as well as to the design of novel synthetic circuits with specified functionality, and a fundamental first step in this direction is to identify the design components of the network that are essential for the *in vivo* physiological response. In the following, we give three examples from the recent systems biology literature of complex cellular control systems whose functionality has been effectively elucidated by this kind of analysis.

1.2.1 The tryptophan operon feedback control system

Tryptophan is one of the 20 standard amino acids, as well as an essential amino acid in the human diet. An operon is a set of structural genes which are closely situated together, functionally related and jointly regulated. Tryptophan is produced through a synthesis pathway from the amino acid precursor chorismate with the help of enzymes which are translated from the tryptophan operon structural genes trpE, D, C, B and A. The dynamics of the tryptophan operon are regulated by an exquisitely complex feedback control system which has been the subject of numerous experimental and modelling studies over recent years (see [11, 12, 13, 14] and references therein).

Transcription of tryptophan is initiated by the binding of the RNA polymerase to the promoter site, and this process is regulated by two feedback mechanisms. The activated aporepressor, which is bound by two molecules of tryptophan, interacts with the operator site and hence represses transcription, [15]. The process of transcription can also be attenuated by binding of the tryptophan molecule to specific mRNA sites. The transcribed mRNA encodes five polypeptides that form the subunits of the enzyme molecules, which in turn catalyse the synthesis of tryptophan from chorismic acid, [14]. The third feedback mechanism results from the binding of the tryptophan molecule to the first enzyme in the tryptophan synthesis pathway, namely anthranilate synthase, thereby inhibiting its activity.

From a control engineering point of view, the tryptophan system in *Escherichia coli* can be conceptualised as a three-processes-in-series system, namely transcription, translation and tryptophan synthesis (P_1, P_2 and P_3 in Fig. 1.5, respectively), [8]. Accurate control of tryptophan concentration in the cell is achieved by three distinct negative feedback controllers, namely genetic regulation, mRNA attenuation and enzyme inhibition (C1, C2 and C3 in Fig. 1.5, respectively). Applications of this kind of parallel or distributed control architecture are widespread both in engineering, [16], and in biological networks. For example, the phosphotases synthesised through high osmolarity glycerol (HOG) activation regulate multiple upstream kinases to modulate the osmotic pressure in *Saccharomyces cerevisiae*, [17]. Another well-known example is the hormonal response in the insulin signalling pathway, [18], in which the phosphorylated Akt and Pkc interact with serially arranged upstream components, namely insulin receptor, insulin receptor substrates and upstream phosphotases, to constitute multiple feedback loops. A similar multiple feedback loop mechanism also exists in p53 regulation of cell-cycle and apoptosis, [19], in which Cdc25 interacts at multiple points of the upstream processes arranged in series. Using this conceptual framework, a number of recent studies have elucidated many of the fundamental design principles of the tryptophan control system, in particular revealing the separate (and non-redundant) roles played by each feedback loop in ensuring a fast and robust response of the tryptophan operon to variations in the level of tryptophan synthesis required by the cell, [13, 20, 8].

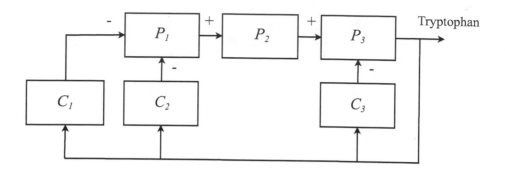

FIGURE 1.5: Tryptophan operon regulatory system in *E. coli* viewed as a distributed feedback control system: tryptophan is the product of three processes in series, namely transcription P_1, translation P_2 and tryptophan synthesis P_3, controlled by three distinct negative feedback controllers, namely genetic regulation C_1, mRNA attenuation C_2 and enzyme inhibition C_3.

1.2.2   The polyamine feedback control system

Polyamines are essential, ubiquitous polycations found in all eukaryotic and most prokaryotic cells. They are utilised in a wide range of core cellular processes such as binding and stabilising RNA and DNA, mRNA translation, ribosome biogenesis, cell proliferation and programmed cell death, [21]. Polyamine depletion results in cell growth arrest [22], whereas over-abundance is cytotoxic, [23, 24]. Thus, homeostatically regulating polyamine content within a relatively narrow non-toxic range is a significant regulatory challenge for the cell.

Ornithine decarboxylase (ODC) is the first and rate limiting enzyme in the biosynthetic pathway which produces the polyamines. The key regulator of ODC in a wide range of eukaryotes is the protein antizyme, [25]. There is a single antizyme isoform in *S. cerevisiae*, Oaz1, [26]. Antizyme binds to and inhibits ODC, and targets it for ubiquitin-independent proteolysis by the 26S proteasome, [27, 28]. Antizyme synthesis is in turn dependent upon a polyamine-stimulated +1 ribosomal frameshift event during translation of its mRNA. Polyamines also inhibit the degradation of antizyme by the ubiquitin pathway, [26]. Polyamines thus regulate their homeostasis via a negative feedback system: a greater concentration of polyamines in the cell increases the rate of antizyme production by stimulating the ribosomal frameshift (and also reduces the rate of antizyme degradation), and a higher concentration of antizyme acts to reduce polyamine levels by increasing the inhibition of ODC.

In control engineering terms, polyamine regulation can be conceptualised

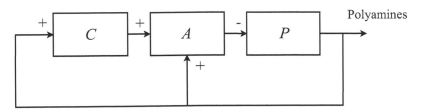

FIGURE 1.6: Polyamine feedback control system in *S. cerevisiae*: a process P (the biosynthetic pathway), affected by an actuator A (the protein antizyme) driven by a feedback controller C (the translational frameshift).

as a process P (the biosynthetic pathway), which is affected by an actuator A (the protein antizyme) under the control of multiple negative feedback loops, Fig. 1.6. In this representation, the translational frameshift event plays the role of a feedback controller C, and a recent study has developed and validated a detailed computational model of the dynamics of this controller in yeast, [29].

To define how each of the polyamines individually and jointly stimulate the frameshifting event at the antizyme frameshift site *in vivo*, a novel quadruple yeast gene knockout strategy was devised in which *de novo* synthesis and metabolic interconversion of supplied polyamines is prevented. Using this experimental tool, this study was able to produce data showing that putrescine, spermidine and spermine stimulate antizyme frameshifting in qualitatively and quantitatively different ways. For example, the effect of putrescine on frameshifting was very weak compared to that of spermidine and spermine, while an analysis of polyamine frameshift responses revealed that although spermidine stimulates frameshifting with a hyperbolic (Michaelis–Menten type) function, in contrast, putrescine and spermine each appear to bind to the ribosome in a cooperative manner.

Using this data, a mathematical function employing both Hill functions and Michaelis–Menten type enzyme kinetics with competing substrates was developed to capture the complex individual and combinatorial effects of the three polyamines on the ribosomal frameshift. This model of the polyamine controller was developed using single and pair-wise polyamine data sets, but was subsequently able to accurately predict frameshift efficiencies measured in both the wild-type strain and in an antizyme mutant, each of which contained very different (triple) polyamine combinations, [29].

1.2.3 The heat shock feedback control system

Cells in living organisms are routinely subjected to stress conditions arising from a variety of sources, including changes in the ambient temperature, the presence of metabolically harmful substances and viral infection. One of the most harmful effects of these types of stress conditions is to cause partial or complete unfolding and denaturation of proteins. Changes in a protein's three-dimensional folded structure will often compromise its ability to function correctly, and as a result widespread unfolding or misfolding of proteins will eventually result in cell death. Natural selection has therefore caused regulatory systems to evolve to detect the damage associated with stress conditions and to initiate a response that increases the resistance of cells to damage and aids in its repair.

One of the most important of these protective systems is the heat shock response, [30], which consists of an elaborate feedback control system which detects the presence of stress-related protein damage, [31], and produces a response to attenuate this disturbance through the synthesis of new heat-shock proteins which can refold denatured cellular proteins, [30, 31, 32, 33]. In *E. coli*, the heat shock response is implemented through a control system centered around the heat shock factor σ^{32}, which regulates the transcription of the heat shock proteins under normal and stress conditions. The enzyme RNA polymerase, bound to the regulatory sigma factor σ^{32}, recognises the heat shock genes that encode molecular chaperones such as DnaK, DnaJ, GroEL, and GrpE, as well as proteases such as Lon and FtsH. Chaperones are responsible for refolding denatured proteins, while proteases degrade unfolded proteins, [32].

The first mechanism through which σ^{32} responds to stress conditions corresponds to an open-loop control system. At low temperatures, the translation start site of σ^{32} is occluded by base pairing with other regions of the σ^{32} mRNA, so that there is little σ^{32} present in the cell and, hence, little transcription of the heat shock genes. When *E. coli* are exposed to high temperatures, this base pairing is destabilised, resulting in a "melting" of the secondary structure of σ^{32}, which enhances ribosome entry, leading to an immediate increase in the translation rate of the mRNA encoding σ^{32}, [34]. Hence, a sudden increase in temperature, sensed through this mechanism, results in a spike of σ^{32} and a corresponding rapid increase in the number of heat shock proteins.

In addition, regulation of σ^{32} is also achieved via two feedback control loops, [32]. The first of these involves the chaperone DnaK and its cochaperone DnaJ. The main function of these chaperones is to perform protein folding, but they can also bind to σ^{32}, therefore limiting the ability of σ^{32} to bind to the RNA polymerase. When the number of unfolded proteins in the cell increases, more of the DnaK/J are occupied with the task of protein folding, and fewer of them are available to bind to σ^{32}. This allows more σ^{32} to bind to RNA polymerase, which in turn causes an increase in the transcription of DnaK/J and other chaperones. The accumulation of high levels of heat shock proteins leads to the efficient refolding of the denatured proteins, thereby decreasing the pool of unfolded protein, and freeing up DnaK/J to again sequester σ^{32} from RNA polymerase. The activity of σ^{32} is thus regulated through a sequestration negative feedback loop that involves competition between σ^{32} and

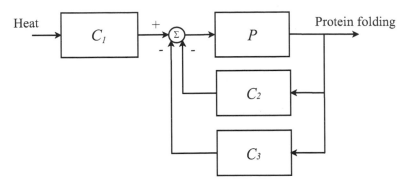

FIGURE 1.7: Heat shock feedback control system in *E. coli*: a process P (transcription and translation of heat shock proteins) controlled by an open-loop feedforward controller C_1 (synthesis of σ^{32}), and two feedback controllers C_2 (sequestration of σ^{32}) and C_3 (degradation of σ^{32}).

the unfolded proteins for binding with the free DnaK/J chaperone pool.

σ^{32} is rapidly degraded ($t_{1/2} = 1$ min) during steady-state growth, but is stabilised for the first five minutes after an increase in temperature. The chaperone DnaK and its cochaperone DnaJ are required for the rapid degradation of σ^{32} by the heat shock protease FtsH. σ^{32} which is bound to RNA polymerase, on the other hand, is protected from this degradation. Furthermore, the synthesis rate of FtsH, a product of the heat shock protein expression, is proportional to the transcription/translation rate of DnaK/J. Therefore, as the number of unfolded proteins increases due to heat shock, the rate of σ^{32} degradation decreases, since fewer DnaK/J are now available in the free chaperone pool and thus more of the σ^{32} will be bound to RNA polymerase. This in turn leads to the production of more DnaJ/K chaperones and more FtsH protease, which brings the σ^{32} degradation rate back up. The activity of σ^{32} is therefore also controlled through a second FtsH degradation negative feedback loop.

After the initial rapid increase in response to heat shock, the concentration of σ^{32} settles to a new steady-state, whose value is determined by the balance between the temperature-dependent positive effects on translation of the σ^{32} mRNA and the negative feedback effects of the heat shock protein chaperones and proteases. The heat shock regulatory system can thus be represented, as shown in Fig. 1.7, as a process P (transcription and translation of heat shock proteins) controlled by an open-loop feedforward controller C_1 (synthesis of σ^{32}), and two feedback controllers C_2 (sequestration of σ^{32}) and C_3 (degradation of σ^{32}). A recent analysis of this system using control engineering methods revealed that the complexity of the hierarchical modular control structures in the heat shock system can be attributed to the necessity of achieving a balance between robustness and performance in the response of the system, [32]. In particular, it was shown that while synthesis of σ^{32} is a powerful strategy that allows for the rapid adaptation to elevated temperatures, it cannot implement a robust response if used by itself in open-loop. On the other hand, the negative feedback loops in the system increase its robustness in the presence of parametric uncertainty and internal fluctuations, but limit the yield for production of heat shock proteins and hence the folding of heat-denatured proteins. Furthermore, the use of degradation feedback implements a faster response to a heat disturbance and reduces the effects of biochemical noise [35, 36].

1.3   Application of control theory to biological systems: A historical perspective

"Engineers have produced many machines that are able to receive and react to information and to exert control by using feedback .... Evidently these machines work very much like living things and we can recognise a great number of feedback systems in the body .... It should be possible to use the precise language developed by the engineers to improve our understanding of those feedback systems that produce the stability of our lives. It cannot be said that physiologists have been able to go very far with this method. The living organism is so complicated that we seldom have enough data to be able to work out exactly what is happening by means of the mathematics the engineer uses. Up to the present, the general ideas and terminology used by these engineers have been of more use to biologists than have the detailed application of their techniques."
J.Z. Young, *Doubt and Certainty in Science : A biologist's reflections on the brain*, The B.B.C. Reith lectures, Oxford University Press, 1950.

As is clear from both the date and content of the above quotation, the idea that control theory could be used to understand the functioning of biological systems is almost as old as control theory itself. Indeed, one of the first books on control theory by a pioneer of the field focussed explicitly on parallels between biological and engineered control systems, [37]. As also noted above, however, difficulties in obtaining sufficient quantities of data meant that for many years the application of control engineering methods in biology would be restricted to the realm of physiology, [38, 39, 40, 41]. Collaborations between control engineers and physiologists led to much fruitful research on systems including the respiratory and cardiovascular systems, [42, 43]; thermoregulation, [40]; water exchange control [39, 44]; blood glucose control, [45]; and pupillary reactions, [46].

The advent over the last decade of high-throughput measurement techniques which can generate -omics level data sets has made possible the simultaneous monitoring of the activity of thousands of genes and the concentrations of proteins and metabolites. This data, and its analysis using sophisticated bioinformatics tools, makes possible for the first time the study of microscopic dynamic interactions among cellular components at a quantitative level. Although much of this data is still of extremely poor quality (when compared to the data on physical systems that control engineers have traditionally had access to), these new measurement technologies have opened the door for the application of control theory to the study of cellular feedback systems, and it is safe to assume that both the quantity and quality of the biological data available will continue to increase over the coming years.

Motivated by the unique dynamic characteristics of cellular systems, control engineers have also begun to develop novel theory which is specifically focussed on biological applications. This situation is clearly an example of a positive feedback loop, in which the successful application of control theory to biological systems spurs the development of new and more powerful theory. As a result, systems biology is rapidly becoming one of the most important application areas for control engineering, as evidenced by the constantly increasing number of sessions dedicated to biology at the major control conferences, and by the number of review papers, special issues of leading control journals and edited volumes on systems biology which have appeared in recent years, [47]-[58]. There seems little doubt that future generations of biologists will collaborate closely with control engineers who are as comfortable dealing with ribosomes and genes as their predecessors were with amplifiers and motors.

References

[1] Rao CV, and Arkin AP. Control motifs for intracellular regulatory networks. *Annual Review Biomedical Engineering*, 3:391–419, 2001.

[2] Doyle J, and Csete M. Motifs, stability, and control. *PLoS Biology*, 3:e392, DOI:10.1371/journal.pcbi.0030392, 2005.

[3] Becskei A, and Serrano L. Engineering stability in gene networks by autoregulation. *Nature*, 405:590–593, DOI:10.1038/35014651, 2000.

[4] Freeman M. Feedback control of intercellular signalling in development. *Nature*, 408:313–319, DOI:10.1038/35042500, 2000.

[5] Goulian M. Robust control in bacterial regulatory circuits. *Current Opinion Microbiology*, 7:198–202, DOI:10.1016/j.mib.2004.02.002, 2004.

[6] Lauffenburger DA. Cell signaling pathways as control mod-

ules: complexity for simplicity?   *PNAS*, 97:5031–5033, DOI:
10.1073/pnas.97.10.5031, 2000.

[7] Yi TM, Huang Y, Simon MI and Doyle JC. Robust perfect adaptation in bacterial chemotaxis through integral feedback control. *PNAS*,
97:4649–4653, DOI:10.1073/pnas.97.9.4649, 2000.

[8] Bhartiya S, Chaudhary N, Venkatesh KV, and Doyle FJ.   Multiple feedback loop design in the tryptophan regulatory network of
Escherichia coli suggests a paradigm for robust regulation of processes in series. *Journal of the Royal Society Interface*, 3:383–391,
DOI:10.1098/rsif.2005.0103, 2006.

[9] Largaespada DA. Haploinsufficiency for tumor suppression: the hazards
of being single and living a long time. *Journal of Experimental Medicine*,
193(4):F15, DOI:10.1084/jem.193.4.F15, 2001.

[10] Kemkemer R, Schrank S, Vogel W, Gruller H, and Kaufmann D. Increased noise as an effect of haploinsufficiency of the tumor-suppressor
gene neurofibromatosis type 1 in vitro. *PNAS*, 99(13):13783–13788,
DOI:10.1073/pnas.212386999, 2002.

[11] Santillan M and Mackey MC. Dynamic regulation of the tryptophan
operon: a modeling study and comparison with experimental data.
PNAS, 98(4):1364–1369, 2001.

[12] Xiu Z-L, Chang ZY, and Zeng A-P. Nonlinear dynamics of regulation of bacterial trp operon: model analysis of integrated effects of repression, feedback inhibition, and attenuation. *Biotechnology Progress*,
18(4):686–693, 2002.

[13] Bhartiya S, Rawool S, and Venkatesh KV Dynamic model of Escherichia
coli tryptophan operon shows an optimal structural design. *European
Journal of Biochemistry*, 270(12):2644–2651, 2003.

[14] Santillan M and Zeron ES. Analytical study of the multiplicity of regulatory mechanisms in the tryptophan operon. *Bulletin of Mathematical
Biology*, 68(2):343–359, 2006.

[15] Yanofsky C and Horn V. Role of regulatory features of the trp operon of
Escherichia coli in mediating a response to a nutritional shift. *Journal
of Bacteriology*, 176(20):6245–6254, 1994.

[16] Balchen JG and Mumme KI. *Process Control Structures and Applications.* New York: Van Nostrand Reinhold, 1988.

[17] Hohmann S.   Osmotic stress signaling and osmoadaptation in
yeasts.   *Microbiology and Molecular Biology Reviews*, 66:300–372,
DOI:10.1128/MMBR.66.2.300-372.2002, 2002.

[18] Sedaghat AR, Sherman A, and Quon MJ. A mathematical model of metabolic insulin signaling pathways. *American Journal of Physiology, Endocrinology and Metabolism*, 283:E1084–E1101, 2002.

[19] Kohn KW. Molecular interaction map of the mammalian cell cycle control and DNA repair systems. *Molecular Biology of the Cell*, 10:2703–2734, 1999.

[20] Venkatesh KV, Bhartiya S, and Ruhela A. Multiple feedback loops are key to a robust dynamic performance of tryptophan regulation in Escherichia coli. *FEBS Letters*, 563:234–240, DOI:10.1016/S0014-5793(04)00310-2, 2004.

[21] Wallace HM, Fraser AV, and Hughes A. A perspective of polyamine metabolism. *Biochemical Journal*, 376:1–14, 2003.

[22] Wallace HM and Fraser AV. Inhibitors of polyamine metabolism. *Amino Acids*, 26:353–365, 2004.

[23] Poulin R, Coward JK, Lakanen JR, and Pegg AE. Enhancement of the spermidine uptake system and lethal effects of spermidine overaccumulation in ornithine decarboxylase-overproducing L1210 cells under hyposmotic stress. *Journal of Biological Chemistry*, 268:4690–4698, 1993.

[24] Tobias KE and Kahana C. Exposure to ornithine results in excessive accumulation of putrescine and apoptotic cell death in ornithine decarboxylase overproducing mouse myeloma cells. *Cell Growth and Differentiation*, 6:1279–1285, 1995.

[25] Ivanov IP and Atkins JF. Ribosomal frameshifting in decoding antizyme mRNAs from yeast and protists to humans: close to 300 cases reveal remarkable diversity despite underlying conservation. *Nucleic Acids Research*, 35:1842–1858, 2007.

[26] Palanimurugan R, Scheel H, Hofmann K, and Dohmen RJ. Polyamines regulate their synthesis by inducing expression and blocking degradation of ODC antizyme. *EMBO Journal*, 23:4857–4867, 2004.

[27] Li X and Coffino P. Regulated degradation of ornithine decarboxylase requires interaction with the polyamine-inducible protein antizyme. *Molecular and Cellular Biology*, 12:3556–3562, 1992.

[28] Murakami Y, Matsufuji S, Kameji T, Hayashi S, Igarashi K, Tamura T, Tanaka K, and Ichihara A. Ornithine decarboxylase is degraded by the 26S proteasome without ubiquitination. *Nature*, 360:597–599, 1992.

[29] Rato C, Amirova SR, Bates DG, Stansfield I, and Wallace HM. Translational recoding as a feedback controller: systems approaches reveal polyamine-specific effects on the antizyme ribosomal frameshift. *Nucleic Acids Research*, 39:4587–4597, DOI: 10.1093/nar/gkq1349, 2011.

[30] Gross C. Function and Regulation of the Heat Shock Proteins, in *Escherichia Coli and Salmonella: Cellular and Molecular Biology.* Washington DC: ASM Press, pp. 1384–1394.

[31] Ang D, Liberek K, Skowyra D, Zylicz M, and Georgopoulos C. Biological role and regulation of the universally conserved heat-shock proteins. *Journal of Biological Chemistry*, 266(36):24233–24236, 1991.

[32] Khammash M and El-Samad H. Systems biology from physiology to gene regulation: engineering approaches to the study of biological sciences. *IEEE Control Systems Magazine*, 62–76, August 2004.

[33] Kurata H, El-Samad H, Iwasaki R, Ohtake H, Doyle JC, Grigorova I, Gross CA, and Khammash M. Module-based analysis of robustness tradeoffs in the heat shock response system. *PLoS Computational Biology*, 2(7):e59, DOI: 10.1371/journal.pcbi.0020059, 2006.

[34] Straus D, Walter W, and Gross C. The activity of σ^{32} is reduced under conditions of excess heat shock protein production in *Escherichia coli*. *Genes and Development*, 3(12A):2003–2010, 1989.

[35] El-Samad H, Kurata H, Doyle JC, Gross CA, and Khammash M. Surviving heat shock: control strategies for robustness and performance. *PNAS*, 102:2736–2741.

[36] El-Samad H and Khammash M. Regulated degradation is a mechanism for suppressing stochastic fluctuations in gene regulatory networks. *Biophysical Journal*, 90:3749–3761.

[37] Wiener N. *Cybernetics, or Control and Communication in the Animal and the Machine.* New York: Wiley, 1948.

[38] Grodins F. *Control Theory and Biological Systems.* New York: Columbia Univ. Press, 1963.

[39] Yamanoto W, and Brobeck J. *Physiological Controls and Regulations.* Philadelphia: Saunders, 1965.

[40] Milsum J, *Biological Control Systems Analysis.* New York: McGraw-Hill, 1966.

[41] Milhorn H, *The Application of Control Theory to Physiological Systems.* Philadelphia: Saunders, 1966.

[42] Khoo MCK. *Physiological Control Systems: Analysis, Simulation, and Estimation.* IEEE Press Series in Biomedical Engineering, 2000.

[43] Batzel JJ, Kappel F, Schneditz D, and Tran HT. *Cardiovascular and Respiratory Systems: Modeling, Analysis, and Control.* SIAM Frontiers in Applied Mathematics, 2006.

[44] Jones RW. *Principles of Biological Regulation: An Introduction to Feedback Systems*. New York: Academic Press, 1973.

[45] Chee F and Tyrone F. *Closed-Loop Control of Blood Glucose*. Lecture Notes in Control and Information Sciences 368. Berlin Heidelberg: Springer, 2007.

[46] Riggs D. *Control Theory and Physiological Feedback Mechanisms*. Baltimore: Williams & Wilkins, 1970.

[47] Special Issue on Biochemical Networks and Cell Regulation. *IEEE Control Systems Magazine*, 24(4), August 2004.

[48] Sontag ED. Some new directions in control theory inspired by systems biology. *IET Systems Biology*, 1:9–18, DOI: 10.1049/sb:20045006, 2004.

[49] Iglesias PA, Khammash M, Munsky B, Sontag ED, and Del Vecchio D. Systems Biology and Control - A Tutorial. In *Proceedings of the 46th IEEE Conference on Decision and Control*, New Orleans, DOI: 10.1109/CDC.2007.4435054, 2007.

[50] Queinnec I, Tarbouriech S, Garcia G, and Niculescu S-I (Editors). *Biology and Control Theory: Current Challenges*. Lecture Notes in Control and Information Sciences 357. Berling Heidelberg: Springer, 2007.

[51] Joint Special Issue on Systems Biology. *IEEE Transactions on Automatic Control* and *IEEE Transactions on Circuits and Systems*, January 2008.

[52] Wellstead P, Bullinger E, Kalamatianos D, Mason O, and Verwoerd M. The role of control and system theory in systems biology. *Annual Reviews in Control*, 32:33–47, 2008.

[53] Wellstead P. On the industrialization of biology. *AI & Society*, 26(1):21–33, DOI: 10.1007/s00146-009-0232-3, 2009.

[54] Iglesias PA and Ingalls BP (Editors). *Control Theory and Systems Biology*. Boston: MIT Press, 2009.

[55] Wellstead P. Systems biology and the spirit of Tustin. *IEEE Control Systems Magazine*, February 2010.

[56] Special Issue on Robustness in Systems Biology: Methods and Applications. *International Journal of Robust and Nonlinear Control*, 20(9):961–1078, June 2010.

[57] Special Issue on System Identification for Biological Systems. *International Journal of Robust and Nonlinear Control*, to appear, 2011.

[58] Special Issue on Systems Biology. *Automatica*, 47(6):1095–1278, 2011.

2

Linear systems

2.1   Introduction

The dynamics of biological systems are generally highly nonlinear — what then is the justification for using linear control system analysis techniques to study such systems? The answer to this question will be familiar to any engineering undergraduate, since it is a fact that while almost *all* real-world control systems display nonlinear dynamics to some extent, the vast majority of the methods used in their design and analysis are based on linear systems theory, from the flight control system on the Airbus A380 to the controller for the servomotor which accesses the hard disk on your computer. Essentially, we have a trade-off: control engineers will, in certain cases, accept the level of approximation involved in modelling the process as a linear system in order to exploit the power, elegance and simplicity of linear analysis and design methods. The key point to remember is that when we model or analyse the dynamics of a particular system we are usually interested only in certain aspects of that system's dynamics — if these may be approximated to a reasonable level of accuracy as a linear system, then there are huge advantages in doing so. The only caveat is that we then need to be careful in interpreting the results of our analysis, as these will hold only within the limitations of the underlying assumptions regarding linearity.

In this chapter, we introduce a number of fundamental techniques for analysing the dynamics of linear systems, and illustrate how they may be used to provide new insight into the design principles of some important biological systems. We discuss the concept of system *state* and *state-space models* before introducing the *frequency response*, which is derived from the *time response* of a *linear time-invariant* (LTI) system to a sinusoidal input. Extending the class of input signals considered to exponential functions of time leads to the concept of *transfer function*, which proves to be a particularly useful tool with which to model and analyse the input–output behaviour of a linear system. The subsequent introduction of the *Fourier* and *Laplace transforms*, along with their related properties, provides the theoretical basis for frequency domain analysis of linear systems. We introduce the notion of *stability* for linear systems, before describing the *characteristic parameters* of the time and frequency response of a linear system. Finally, we show how interconnected systems may be conveniently represented using block diagrams, as per standard practice in control engineering.

2.2   State-space models

A state-space representation is a mathematical model of a system as a set of input, output and state variables related by first-order differential equations. The state variables of a system can be regarded as a set of variables that uniquely identifies its current condition. For mechanical systems, typical state variables include values of the system's position and velocity, while for thermodynamic systems the states may include temperature, pressure, entropy, enthalpy and internal energy. In systems biology, state variables are typically just the concentrations of the different molecular species which are changing over time, e.g. mRNA, proteins, metabolites, ligands, receptors, etc. Given a state-space model, the knowledge of the state at time t_0 allows the computation of the system's evolution for all $t > t_0$, even in the absence of any information about the inputs at time $t < t_0$.

State-space models are commonly used in control engineering because they provide a convenient and compact way to model and analyse high-order systems* with multiple inputs and outputs. Also, unlike the frequency domain representations to be introduced later in this chapter, the use of state-space models is not limited to systems with linear components and zero initial conditions. Consider the following differential equation model of a system with input $u(t)$ and output $y(t)^{\dagger}$:

$$\frac{d^3y(t)}{dt^3} + \frac{5d^2y(t)}{dt^2} + \frac{3dy(t)}{dt} + 4y(t) = u(t) \qquad (2.1)$$

To write this system as a state-space model, we must first define the state variables for the system. The minimum number of state variables required to model a system is equal to the order of the corresponding differential equation model. In this case, therefore, we have three state variables given by

$$x_1 = y$$
$$x_2 = \dot{y} = \dot{x}_1$$
$$x_3 = \ddot{y} = \dot{x}_2$$

---

*Systems whose governing equations involve derivatives of high order.
†Note that in the future, for convenience, we usually drop the explicit dependence of inputs, outputs and states on time in the notation.

In terms of the state variables, we can write the original differential equation model (Eq. 2.1) as:

$$\dot{x}_1 = x_2$$
$$\dot{x}_2 = x_3$$
$$\dot{x}_3 = -4x_1 - 3x_2 - 5x_3 + u$$

Note that we have thus converted a third-order differential equation model into a model consisting of three first-order differential equations. One great advantage of this formulation is that we can use matrix/vector notation to represent the system in a highly compact form. Writing

$$\begin{pmatrix} \dot{x}_1 \\ \dot{x}_2 \\ \dot{x}_3 \end{pmatrix} = \begin{pmatrix} 0 & 1 & 0 \\ 0 & 0 & 1 \\ -4 & -3 & -5 \end{pmatrix} \begin{pmatrix} x_1 \\ x_2 \\ x_3 \end{pmatrix} + \begin{pmatrix} 0 \\ 0 \\ 1 \end{pmatrix} u$$

$$y = \begin{pmatrix} 1 & 0 & 0 \end{pmatrix} \begin{pmatrix} x_1 \\ x_2 \\ x_3 \end{pmatrix}$$

if we now define the *state vector* x as

$$x = \begin{pmatrix} x_1 \\ x_2 \\ x_3 \end{pmatrix}$$

then the state-space model can be simply written as

$$\dot{x} = Ax + Bu \qquad (2.2a)$$
$$y = Cx + Du \qquad (2.2b)$$

The state-space model is thus completely defined by specifying the state vector x and the values of the four matrices A, B, C and D, which in this case are given by:

$$A = \begin{pmatrix} 0 & 1 & 0 \\ 0 & 0 & 1 \\ -4 & -3 & -5 \end{pmatrix}, \quad B = \begin{pmatrix} 0 \\ 0 \\ 1 \end{pmatrix}, \quad C = \begin{pmatrix} 1 & 0 & 0 \end{pmatrix}, \quad D = 0 \qquad (2.3)$$

Note that the set of state variables is not unique: infinitely many state-space representations can be generated for a given system by applying the linear transformation $z = Tx$ (where z is the new state vector and T is invertible) and these representations are all equivalent in terms of their input–output behaviour. Another great advantage of state-space models is that they can easily represent both single-input single-output (SISO) and multi-input multi-output (MIMO) systems. Consider a system with two inputs u_1 and u_2 and two outputs y_1 and y_2, whose dynamics are given by the differential equations

$$\frac{d^3y_1}{dt^3} + \frac{5d^2y_1}{dt^2} + \frac{3dy_1}{dt} + 4y_1 = u_1$$

$$\frac{d^3y_2}{dt^3} + \frac{d^2y_1}{dt^2} + \frac{4dy_2}{dt} + 2(y_1 + y_2) = u_2$$

The state variables of this system are given by

$$x_1 = y_1 \qquad\qquad\qquad x_4 = y_2$$
$$x_2 = \dot{y}_1 = \dot{x}_1 \qquad\qquad x_5 = \dot{y}_2 = \dot{x}_4$$
$$x_3 = \ddot{y} = \dot{x}_2 \qquad\qquad x_6 = \ddot{y}_2 = \dot{x}_5$$

If we now define the input, output and state vectors for this system as

$$u = \begin{pmatrix} u_1 \\ u_2 \end{pmatrix}, \quad y = \begin{pmatrix} y_1 \\ y_2 \end{pmatrix}, \quad x = \begin{pmatrix} x_1 \\ x_2 \\ x_3 \\ x_4 \\ x_5 \\ x_6 \end{pmatrix}$$

then the state-space model can again be simply written in the form of Eq. (2.2), where now we have

$$A = \begin{pmatrix} 0 & 1 & 0 & 0 & 0 & 0 \\ 0 & 0 & 1 & 0 & 0 & 0 \\ -4 & -3 & -5 & 0 & 0 & 0 \\ 0 & 0 & 0 & 0 & 1 & 0 \\ 0 & 0 & 0 & 0 & 0 & 1 \\ -2 & 0 & -1 & -2 & -4 & 0 \end{pmatrix}, \quad B = \begin{pmatrix} 0 & 0 \\ 0 & 0 \\ 1 & 0 \\ 0 & 0 \\ 0 & 0 \\ 0 & 1 \end{pmatrix},$$

$$C = \begin{pmatrix} 1 & 0 & 0 & 0 & 0 & 0 \\ 0 & 0 & 0 & 1 & 0 & 0 \end{pmatrix}, \quad D = \begin{pmatrix} 0 & 0 \\ 0 & 0 \end{pmatrix}. \qquad (2.4)$$

2.3   Linear time-invariant systems and the frequency response

It may seem entirely natural to describe dynamical systems, whose inputs, outputs and states vary with time, using time domain models such as differential equations or state-space representations. In control engineering, however, it has long been recognised that the time domain approach is sometimes not ideal for the analysis of signals and systems, due to the inherent complexity of the theoretical and numerical machinery needed to deal with differential equation-based problems. Moreover, in many applications (e.g. telecommunications) one is more interested in the harmonic content of a signal, and how this is modified when it is passed through a given system (in this case the system is usually termed a *filter*), rather than the temporal evolution of the signal. Likewise, for biological systems where the processing of information through cellular signalling cascades may occur over a wide range of time scales (e.g. fast ligand-receptor dynamics versus the much slower response of gene expression changes), analysing and characterising the dynamics of a system in the frequency domain can provide deep insight into the dominant processes which dictate the overall response of the system.

The class of systems that will be considered in this chapter is characterised by the linearity of the response and by the fact that the system dynamics do not change over time. More precisely, given a system with input $u(t)$ and output $y(t, u(t))$, we say that it is *linear* if the following condition is satisfied:

$$y\left(t, \alpha u_1(t) + \beta u_2(t)\right) = \alpha y\left(t, u_1(t)\right) + \beta y\left(t, u_2(t)\right), \quad \forall \alpha, \beta \in \mathbb{R}. \quad (2.5)$$

The above condition states that a linear combination of two (or more) inputs yields a linear combination of the corresponding outputs with the same coefficients. This property is usually referred to as the *Superposition Principle*. It is straightforward to recognise that Eq. (2.5) implies

$$y\left(t, \alpha u(t)\right) = \alpha y\left(t, u(t)\right), \quad \forall \alpha \in \mathbb{R},$$

that is, scaling the input by α produces a scaling of the output by the same factor.

Note that, in the general case, the system response y explicitly depends on the time (in this case the system is said to be *time–varying*): this entails that, if we subject the system to two identical inputs at two different points in time, the outputs will be different. A *time-invariant* system, instead, has the nice property of always producing the same output when subject to the same input, independently of the time at which the input is applied. This property can be stated mathematically as

$$y\left(t_1, u(t)\right) = y\left(t_2, u(t)\right), \quad \forall t_1, t_2 \in \mathbb{R}. \quad (2.6)$$

Let us consider a Linear Time-Invariant (LTI) system with input u and output y. Under certain conditions[‡], if we subject this system to a sinusoidal input signal

$$u(t) = A \sin(\omega t + \theta),$$

[‡]The system must be asymptotically stable; the concept of stability is formally defined in Section 2.6.

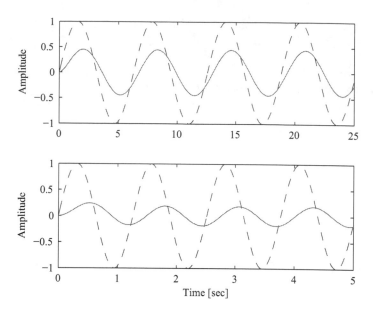

FIGURE 2.1: Outputs (solid lines) of an LTI system subject to two sinusoidal inputs (dashed lines) of different frequencies.

with amplitude A, frequency ω (in rad/s) and phase θ (in rad), after an initial transient we will obtain a steady state sinusoidal output signal with the same frequency

$$y_\infty(t) = M(\omega)A\sin(\omega t + \theta + \varphi(\omega)). \qquad (2.7)$$

It is important to realise that this property does not hold in the general case (i.e. when the system is nonlinear and/or time-varying). Note also that the amplitude scaling factor and the additional phase term are functions of the frequency of the sinusoidal input; therefore, we can define the *frequency response function*

$$H(\omega) : \omega \in \mathbb{R} \mapsto M(\omega)e^{i\varphi(\omega)} \in \mathbb{C}. \qquad (2.8)$$

Example 2.1

Let us consider the simple first-order model

$$\dot{y} + 2y = u.$$

Fig. 2.1 shows the response of the system to the sinusoidal inputs

$$u_1(t) = \sin(t), \quad u_2(t) = \sin(5t).$$

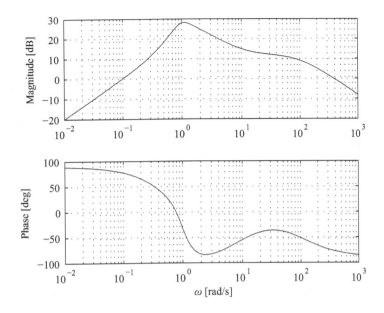

FIGURE 2.2: Representation of a frequency response through Bode plots.

Although both inputs have unit amplitude and null phase, the response of the system to u_1 (upper subplot) exhibits a greater amplitude and a smaller phase lag than the response to u_2 (lower subplot). □

The standard form used for representing the frequency response of a system in control engineering is through Bode plots. Bode plots consist of two diagrams, which give the values of the magnitude and phase of $H(\omega)$ as a function of ω. A logarithmic (base 10) scale is used on the ω-axis, since one is typically interested in visualising with uniform precision the behaviour of the system over a wide range of frequencies. The magnitude or *gain* of the system is given in decibels, computed as

$$|H(\omega)|_{\mathrm{dB}} = 20 \log_{10} |H(\omega)|,$$

whereas a linear scale is used for the phase $\varphi(\omega)$, which is measured in degrees. Fig. 2.2 shows an example of a Bode plot for an LTI system: the frequency response is plotted over the range of frequencies $\omega \in \left[10^{-2}, 10^3\right]$. Correspondingly, the gain varies between -25 and 25 db (0.00316 to 316 on a linear scale, i.e. five orders of magnitude). Note that the Bode plots are displayed only for positive frequency values and the point $\omega = 0$ is located at $-\infty$ on the horizontal axis; moreover, the gain is always positive (negative dB values correspond to magnitudes less than unity). Note also that, since the gain is given on a logarithmic scale, the plot of the product of two different functions of ω

TABLE 2.1
Common input signals

Constant	$k = ke^{0t}$
Real-valued exponential	$ke^{\alpha t}$
Sinusoid	$\sin(\omega t) = 0.5 \left(e^{i\omega t} - e^{-i\omega t}\right)$
Growing/decaying sinusoid	$e^{\alpha t}\sin(\omega t) = 0.5 \left(e^{(\alpha+i\omega)t} - e^{(\alpha-i\omega)t}\right)$

may be obtained by just summing their values on their respective Bode plots at each frequency, that is

$$|H_1(\omega)H_2(\omega)|_{\mathrm{dB}} = 20\log_{10}(|H_1(\omega)||H_2(\omega)|)$$
$$= 20\log_{10}(|H_1(\omega)|) + 20\log_{10}(|H_2(\omega)|)$$
$$= |H_1(\omega)|_{\mathrm{dB}} + |H_2(\omega)|_{\mathrm{dB}}$$

Finally, we define a *decade* on the x-axis of a Bode plot as the interval defined by two frequency values which differ by a factor of 10 (e.g. [2, 20]).

If we now consider input signals belonging to the class of complex exponential functions, that is

$$u(t) = e^{st}, \quad s = \alpha + i\omega \in \mathbb{C},$$

the arguments above can be generalised to a broader class of signals. To understand the practical usefulness of complex functions, recall that

$$e^{(\alpha+i\omega)t} = e^{\alpha t}e^{i\omega t} = e^{\alpha t}\left(\cos(\omega t) + i\sin(\omega t)\right).$$

Many types of real-valued input signals may be written as a linear combination of complex exponential functions by exploiting the above relations (see Table 2.1). Therefore, if we know how to compute the response of an LTI system to a complex exponential input, the response to a wide variety of signals can be readily derived by applying the *superposition principle*.

Avoiding the mathematical derivation (which can be found in any linear systems textbook), we can state that the response of an LTI system to an exponential input[§], $u(t) = e^{st}$, takes the form

$$y(t) = \tilde{y}(t) + y_\infty(t)$$
$$= Ce^{At}\left(x(0) - (sI - A)^{-1}B\right) + \left(C(sI - A)^{-1}B + D\right)e^{st}, \qquad (2.9)$$

where A, B, C, D are the matrices of a state-space model of the system in the form of Eq. (2.2), with state vector $x(t)$, for all values of s, except those corresponding to the eigenvalues of A (see below). The first term, $\tilde{y}(t)$, which

---

[§]For notational convenience, we consider only values of $t \geq 0$ and assume that the input is applied at time $t = 0$.

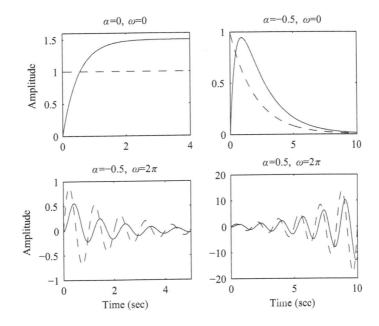

FIGURE 2.3: Outputs (solid lines) of an LTI system subject to different exponential inputs (dashed lines).

is proportional to the matrix-valued exponential e^{At}, is denoted the *transient response* of the system to signify that, in those cases when $e^{At} \to 0$ as $t \to \infty$, this term eventually converges to zero. The second term, $y_\infty(t)$, denoted the *steady state response*, is independent of x_0 and proportional to the input and thus exhibits the same exponential form. Note that in general it is possible to find an initial condition x_0 that nullifies the transient response, yielding only the steady state response, i.e. the response remaining after the initial transient has died away.

Example 2.2
   In order to illustrate the property of LTI systems subject to exponential inputs, we compute the responses of the system

$$\dot{y}(t) + 2y(t) = 3u(t) \tag{2.10}$$

to several inputs, namely a constant signal, a decaying exponential, and decaying and growing sinusoids (see Table 2.1). The signals are assumed to be null for $t < 0$ and the initial conditions of the ODE are zero, that is $y(0) = 0$, $\dot{y}(0) = 0$. In all of these cases we can note (see Fig. 2.3) that the output, after an initial transient, assumes the same shape as the input.   □

From the above discussion, it is possible to define the *Transfer Function*

$$G(s) : s \in \mathbb{C} \mapsto C\,(sI - A)^{-1} + D \in \mathbb{C}, \tag{2.11}$$

of a system which maps the *generalised frequency* s to the steady state response of an LTI system to the input e^{st}. The transfer function can also be defined as the ratio of the steady state output signal and the exponential input signal, that is

$$G(s) = \frac{y_\infty(t)}{e^{st}}.$$

It is important to remark that the arguments above only hold when $s \neq \lambda_j(A)$, $j = 1, \dots, n$, the eigenvalues of A; indeed, this guarantees that $(sI - A)$ is nonsingular and can be inverted. In the case $s = \lambda_j(A)$, the response takes the form

$$y_\infty(t) = \left(c_0 t^r + \cdots + c_{r-1} t + c_r\right) e^{\lambda_j t},$$

where $r + 1$ is the algebraic multiplicity¶ of the eigenvalue λ_j, and $c_0, \dots, c_r$ are constants.

The transfer function is closely connected to the frequency response in Eq. (2.8); indeed, the latter can be obtained from the former by restricting s to belong to the imaginary axis

$$H(\omega) = G(s)|_{s=i\omega}, \tag{2.12}$$

under the hypothesis that $\Re(\lambda_j) < 0, \forall j = 1, \dots, n$.

2.4   Fourier analysis

The results given in the previous section for sinusoidal inputs constitute the basis for a more general treatment of the input–output behaviour of linear systems. This generalisation is based on the fact that the vast majority of signals of practical interest can be written as the sum of a (finite or infinite) set of sinusoidal terms, named the harmonics of the signal. This fact, along with the assumption of linearity (which implies the superposition of multiple effects), enables us to interpret the time response of a system in terms of its frequency response, that is the sum of the responses to each harmonic of the input signal.

¶An eigenvalue with algebraic multiplicity r appears r times as a root of the characteristic polynomial of the system — see Section 2.5.

A fundamental result in Fourier analysis is that any periodic function $f(t)$ with period T can be written as

$$f(t) = F_0 + \sum_{n=1}^{\infty} [F_{cn} \cos(n\omega_0 t) + F_{sn} \sin(n\omega_0 t)] ,$$

where $\omega_0 = 2\pi/T$, and

$$F_0 = \frac{1}{T} \int_T f(t) dt ,$$

$$F_{cn} = \frac{2}{T} \int_T f(t) \cos(n\omega_0 t) dt , \quad F_{sn} = \frac{2}{T} \int_T f(t) \sin(n\omega_0 t) .$$

Note that F_0 is the average value of $f(t)$ over a single period.

Example 2.3

Consider the square wave signal

$$P_w(t) = \begin{cases} 1 & \text{if} \quad 0 < t \le T/2 \\ 0 & \text{if} \quad T/2 < t \le T \end{cases}$$

Using the formula above we can easily compute the Fourier coefficients

$$F_0 = \frac{1}{2}, \quad F_{cn} = 0 \quad \forall n \in \mathbb{N}, \quad F_{sn} = \begin{cases} \frac{2}{n\pi} & \text{if } n \text{ is odd} \\ 0 & \text{if } n \text{ is even} \end{cases} .$$

Therefore, the square wave can be written

$$P_w(t) = \frac{1}{2} + \frac{2}{\pi} \sin(\omega_0 t) + \frac{2}{3\pi} \sin(3\omega_0 t) + \frac{2}{5\pi} \sin(5\omega_0 t) + \cdots \qquad (2.13)$$

Fig. 2.4 shows different approximations of a square wave signal, obtained using the average value plus an increasing number of harmonics. ☐

The previous example shows that a periodic signal can often be well approximated by the sum of a small number of sinusoidal terms. However, if such a periodic signal is used as an input to an LTI system, then even the use of more than one or two harmonics might be redundant: since the output is made up of the sum of the responses to each sinusoidal term and the magnitude scaling factor at the frequency $n\omega_0$ is usually decreasing as n increases[||], in the sum we can in practice often neglect the terms with high n. In the general case, in order to compute a good approximation we have to consider both the Fourier coefficients and the *bandwidth* of the frequency response of the system, which will be defined in Section 2.9.

[||] Furthermore, many systems exhibit a so-called low-pass frequency response, that is the magnitude of $H(\omega)$ decreases when $\omega \to \infty$.

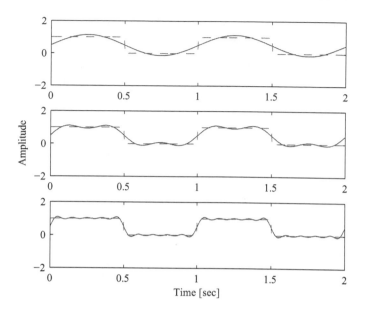

FIGURE 2.4: Approximation of a square wave with the average value plus (a) one harmonic, (b) two harmonics, (c) five harmonics.

Example 2.4

Let us consider the system whose transfer function is

$$G(s) = \frac{1}{s^2 + s + 1} \tag{2.14}$$

and assume we want to compute its steady state response to the square wave $P_w(t)$ with period $T = 2\pi$. In theory, we could compute it as the sum of the steady state responses to each of the terms appearing in Eq. (2.13). In practice, looking at the amplitude of the Fourier coefficients of the harmonics and the magnitude frequency response of the system shown in Fig. 2.5, we can readily recognise that the amplitude of the response associated with the first harmonic will be much greater than the others, thus dominating the overall response. Indeed, the amplitude of each harmonic of the output signal can be computed by summing the two graphs at each frequency. For example, the second harmonic at 3 rad/s has one third the amplitude of the first harmonic; moreover, the value of $M(\omega)$ at 3 rad/s is about one tenth of its value at 1 rad/s. Hence, in the output signal, the amplitude of the second harmonic will be 1/30 of the first harmonic and higher harmonics will be attenuated even further.

To confirm the result of the frequency domain analysis, in Fig. 2.6 we show that the outputs of system (2.14) subject to the square wave and with just

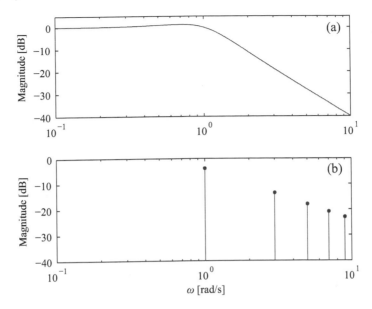

FIGURE 2.5: (a) Magnitude frequency response of system (2.14) and (b) magnitude of the first five Fourier coefficients of the square wave input.

the first two terms of the Fourier expansion (2.13) are practically identical.
☐

As we will see later, an important consequence of the above result is that the frequency response of a system, i.e. the response of a system to sinusoidal signals of different frequencies, can often be approximately evaluated by applying square wave inputs, which are much easier to produce in typical experimental conditions.

So far in this section, we have assumed that the signals whose effect on a system we wish to analyse are periodic. In this case, the frequency spectrum (the coefficients of the Fourier series) of the signal is discrete (i.e. it is defined only at certain frequencies). When the signal is aperiodic, we can think of it as a periodic signal with period $T = \infty$. Thus, the interval between two consecutive harmonics $n\omega_0 = n2\pi/T$ tends to zero and the frequency spectrum becomes a continuous function of ω (i.e. defined for all frequency values). Formally, given an aperiodic signal $f(t)$, it can be analysed in the frequency domain by applying the Fourier transform, defined as

$$\mathcal{F}(\omega) = \int_{-\infty}^{\infty} f(t)e^{-j\omega t}dt. \qquad (2.15)$$

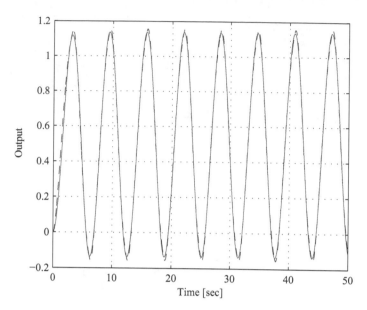

FIGURE 2.6: Outputs of system (2.14) subject to the square wave (dashed) and with the first harmonic approximation (solid) signals shown in Fig. 2.4(a).

2.5   Transfer functions and the Laplace transform

Given an LTI system, described by an input–output ODE model in the form

$$y^{(n)}(t) + a_1 y^{(n-1)}(t) + \cdots + a_n y(t) =$$
$$= b_0 u^{(n)}(t) + b_1 u^{(n-1)}(t) + \cdots + b_m u(t), \qquad (2.16)$$

the computation of the corresponding transfer function is straightforward. Indeed, as discussed above, if we apply the input $u(t) = e^{st}$ at $t = 0$, the steady state response takes the form $y_\infty(t) = y_0 e^{st}$. By substituting into Eq. (2.16) we get

$$\left(s^n + a_1 s^{n-1} + \cdots + a_n\right) y_0 e^{st} = \left(b_0 s^n + b_1 s^{n-1} + \cdots + b_n\right) e^{st},$$

which yields

$$G(s) = \frac{y_\infty(t)}{u(t)} = \frac{b_0 s^n + b_1 s^{n-1} + \cdots + b_n}{s^n + a_1 s^{n-1} + \cdots + a_n} = \frac{N(s)}{D(s)}. \qquad (2.17)$$

Note that we have implicitly assumed $D(s) \neq 0$. $D(s)$ is the characteristic polynomial of the ODE (2.16) and its roots coincide with the eigenvalues of the matrix A in Eq. (2.9); hence $D(s) \neq 0$ is equivalent to $s \neq \lambda_j$. Such values are also named the *poles* of the transfer functions, whereas the roots of the numerator of Eq. (2.17) are the *zeros*. The number n is the *order* of the system; therefore a system of order n has n poles, which can assume either real or complex conjugate values[**].

The transfer function can be exploited not only for computing the steady state response to exponential inputs, but also the response to a generic signal (assuming zero initial conditions, i.e. $x(0) = 0$ in Eq. (2.9)). In order to illustrate this, it is convenient to introduce a particular operator, which enables the transformation of signals from the time domain to the generalised frequency domain (or s-domain). Given a real-valued function $f(t)$, the *Laplace Transform*, $\mathcal{L}$, maps $f(t)$ to a complex-valued function $F(s)$

$$f : t \in \mathbb{R}^+ \mapsto f(t) \in \mathbb{R} \quad \overset{\mathcal{L}}{\underset{\mathcal{L}^{-1}}{\rightleftharpoons}} \quad F : s \in \mathbb{C} \mapsto F(s) \in \mathbb{C} \qquad (2.18)$$

through the relation

$$F(s) = \int_0^\infty f(t)e^{-st} dt . \qquad (2.19)$$

To be rigorous, we must say that the Laplace transform can be applied only if the function f exhibits certain mathematical properties, although practically speaking these are almost always satisfied by the signals commonly of interest in real-world applications. Note also that f is defined only for positive time values. Despite its apparent complexity, the practical application of the Laplace transform is straightforward, due to some special properties:

a) The Laplace transform is linear, i.e. if we consider two functions, $f(t)$ and $g(t)$, we get

$$\mathcal{L}\left(k_1 f(t) + k_2 g(t)\right) = k_1 \mathcal{L} f(t) + k_2 \mathcal{L} g(t) .$$

b) The derivative operator with respect to time corresponds to a multiplication by s in the s-domain

$$\mathcal{L}\frac{df(t)}{dt} = s\mathcal{L} f(t)$$

c) Analogously, the integral operator with respect to time corresponds to a division by s in the s-domain

$$\mathcal{L}\int_0^t f(\tau)d\tau = \frac{1}{s}\mathcal{L} f(t) - f(0) .$$

[**]This stems from the fact that the denominator $D(s)$ is a polynomial of order n with real-valued coefficients.

These and other properties allow us to readily transform differential equations in the time domain into algebraic equations in the s-domain using the Laplace transform. The advantage is evident: it is much easier to solve algebraic equations than differential ones. Once we have found the solution in the s-domain, we can obtain the time domain solution by applying the inverse Laplace transform (or antitransform), $\mathcal{L}^{-1}$. In practice, signals can be easily transformed and antitransformed, without performing any involved calculations, by using readily available tables of Laplace transforms (see, for example, page 799 of [1]).

At this point, we can introduce an equivalent alternative definition of the transfer function of a system as the ratio of the Laplace transforms of the output and input signals, i.e. for an LTI system with zero initial conditions

$$G(s) = \frac{Y(s)}{U(s)} = \frac{\mathcal{L}y(t)}{\mathcal{L}u(t)}. \tag{2.20}$$

Note that the input signal can assume any form (not only exponential) and y denotes the total response (not only the steady state term).

Example 2.5

Assume we want to compute the response of the system (2.10) to the input $u(t) = (2 + cos(10t))$ applied at time $t = 0$. Instead of solving the differential equation, we compute the transfer function (using Eq. (2.16)–(2.17))

$$G(s) = \frac{3}{s+2}$$

and the Laplace transform of the input

$$U(s) = \frac{2}{s} + \frac{s}{s^2 + 100},$$

to derive the Laplace transform of the output

$$Y(s) = G(s)U(s) = \frac{9\,s^2 + 600}{s\,(s^2 + 100)\,(s+2)}.$$

The time response can then be computed by first decomposing $Y(s)$ into a sum of elementary fractional terms, of the same type as those appearing in standard Laplace transform tables, that is

$$Y(s) = \frac{3}{s} - \frac{3.058}{s+2} + \frac{0.0577s + 2.885}{s^2 + 100}.$$

Exploiting the linearity of the Laplace transform, we can now readily derive the time response as the sum of the antitransform of each term

$$y(t) = \left[3 - 3.058\,e^{-2t} + 0.294 \cos\left(10\,t - 1.373\right)\right] 1(t).$$

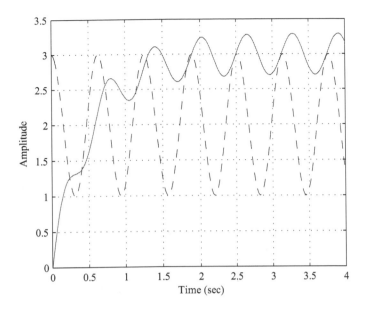

FIGURE 2.7: Time response (solid line) of system (2.10) to the input $u(t) = (2 + cos(10t))$ (dashed line).

The time courses of the input and output are shown in Fig. 2.7. The steady state response is the sum of the first and third terms; indeed, the second term goes to zero as $t \to \infty$. The same result can be found by exploiting the relations illustrated in Section 2.3, and individually computing the steady state responses to the two terms composing the input, that is $u_1(t) = 2 \cdot 1(t)$ and $u_2(t) = cos(10t)1(t)$, which yields

$$y_{1\infty} = 2\,G(i0) = 3\,,$$

$$y_{2\infty} = |G(i10)|\cos\left(10\,t + \angle G(i10)\right) = 0.294\cos\left(10\,t - 1.373\right).$$

□

2.6 Stability

The previous example highlights the fact that the response of an LTI system heavily depends on the value of the poles of the transfer function. In particular, the sign of the real part of the poles determines the sign of the exponent of the exponential terms which, in turn, yield a convergent (if they are all negative) or divergent (if there is at least one positive) response. More precisely, we will distinguish between three classes of systems, namely

- *Asymptotically Stable*: the system poles all have negative real parts. A bounded input (e.g. a step) will produce a bounded output; moreover the output will asymptotically (i.e. for $t \to \infty$) tend to zero when the input is set back to zero.

- *Stable*: the poles all have nonpositive real parts, or there is at most one pole at the origin, or there is at most one pair of poles on the imaginary axis. For this class of systems, bounded inputs produce bounded outputs; however, if the input is set back to zero, the output does not necessarily converge to zero.

- *Unstable*: there is at least one pole with a positive real part or at least two poles at the origin, or there are at least two pairs of poles on the imaginary axis. The output of the system can diverge to infinity even when subject to a bounded input signal.

The stability of a linear system may also be evaluated by computing the eigenvalues of the A matrix (known as the state transition matrix) in the system's state-space model. To see this, we first make explicit the relation between the state-space representation of a single-input single-output (SISO) linear system and its transfer function. Starting with the state-space model

$$\dot{x} = Ax + Bu \tag{2.21a}$$
$$y = Cx + Du \tag{2.21b}$$

and assuming zero initial conditions, taking the Laplace transform gives

$$sX(s) = AX(s) + BU(s) \tag{2.22a}$$
$$Y(s) = CX(s) + DU(s) \tag{2.22b}$$

Solving for $X(s)$ in the first of these equations gives

$$(sI - A)X(s) = BU(s) \tag{2.23}$$

or

$$X(s) = (sI - A)^{-1}BU(s) \tag{2.24}$$

where I is the identity matrix. Now substituting Eq. 2.24 into Eq. 2.22b gives

$$Y(s) = C(sI - A)^{-1}BU(s) + DU(s)$$
$$= \left[C(sI - A)^{-1}B + D\right]U(s) \tag{2.25}$$

Thus, the transfer function $G(s)$ of the system may be defined in terms of the state-space matrices as

$$G(s) = \frac{Y(s)}{U(s)} = C(sI - A)^{-1}B + D$$

$$= C\frac{\text{adj}(sI - A)}{\det(sI - A)}B + D \tag{2.26}$$

Now, recall that the eigenvalues of a matrix A are those values of λ that permit a nontrivial ($x \neq 0$) solution for x in the equation

$$Ax = \lambda x \qquad (2.27)$$

Writing the above equation as

$$(\lambda I - A)x = 0 \qquad (2.28)$$

and solving for x yields

$$x = (\lambda I - A)^{-1} 0 \qquad (2.29)$$

or

$$x = \frac{\text{adj}(\lambda I - A)}{\det(\lambda I - A)} 0 \qquad (2.30)$$

Thus, for a nontrivial solution for x we require that

$$\det(\lambda I - A) = 0 \qquad (2.31)$$

and the corresponding values of λ are the eigenvalues of the matrix A. Comparing Eq. 2.31 with Eq. 2.26, we can now see clearly that the eigenvalues of the state transition matrix A are identical to the poles of the system's transfer function. Noting that all of the above development also holds for multiple-input multiple-output (MIMO) systems, we see that checking the stability of a system represented in state-space form simply requires us to check the location of the eigenvalues of the system's A matrix on the s-plane.

Note carefully that all of the above results hold only in the case of LTI systems — both the definition and analysis of stability are significantly more complex in the case of nonlinear systems, as will be discussed in detail in the next chapter.

2.7 Change of state variables and canonical representations

For LTI systems, input–output representations in the time and frequency domain can be put in a one to one correspondence in view of Eq. (2.16)–(2.17). In the case of input-state-output (ISO) representations, in the previous section we have seen how the transfer function of an LTI system can be readily derived from a state-space model. The question naturally arises as to whether it is possible to derive a state-space model from an assigned transfer function, which is the problem known as *realisation* in systems theory.

Recall that the set of state variables is not uniquely determined: if we consider a generic state-space model in the form of Eq. (2.21a), infinitely many other state-space representations can be generated for the same system by changing the state variables using the linear transformation $z = Tx$ (where z is the new state vector and T is invertible). These alternative representations are all equivalent in terms of their input–output behaviour: applying the transformation to system (2.21a), for example, we obtain the generic transformed system

$$\dot{z} = TAT^{-1}z + TBu \tag{2.32a}$$

$$y = CT^{-1}z + Du \tag{2.32b}$$

and, computing the transfer function of the transformed system, that is

$$CT^{-1}\left(sI - TAT^{-1}\right)^{-1}TB + D = C\left(sI - A\right)^{-1}B + D\,, \tag{2.33}$$

we see that it is identical to that of the original system.

Therefore, given a certain transfer function, it is not possible to uniquely derive one equivalent state-space representation. However, among the infinite possible state-space representations there are some that have a special structure that greatly simplifies the realisation problem. One of these *canonical state-space forms* is known as the *observability form*, which transforms the transfer function (2.17) (or the IO time domain model (2.16)) to the state-space model

$$\dot{x} = \begin{pmatrix} 0 & 0 & 0 & \ldots & 0 & -a_0 \\ 1 & 0 & 0 & \ldots & 0 & -a_1 \\ 0 & 1 & 0 & \ldots & 0 & -a_2 \\ \vdots & \vdots & & \ddots & \vdots & \vdots \\ 0 & 0 & 0 & \ldots & 1 & -a_{n-1} \end{pmatrix} x + \begin{pmatrix} \hat{b}_0 \\ \hat{b}_1 \\ \hat{b}_2 \\ \vdots \\ \hat{b}_{n-1} \end{pmatrix} u \tag{2.34a}$$

$$y = \begin{pmatrix} 0 & 0 & 0 & \ldots & 0 & 1 \end{pmatrix} x + \hat{b}_n u \tag{2.34b}$$

with

$$\hat{b}_n = b_n, \quad \hat{b}_i = b_i - a_i b_n, \quad i = 0, \ldots, n-1\,.$$

As we shall see in Case Study I, using this canonical form can greatly facilitate the analysis of the internal dynamics of a system, since the measurable output can then be assumed to coincide with one of the state variables.

2.8   Characterising system dynamics in the time domain

In this section, we give a brief description of some of the most important measures used in control engineering for characterising time domain dynamics, since these measures can be very useful in evaluating and comparing the performance of different biological control systems.

The dynamics of a control system may be characterised by considering the nature of its response to particular inputs. Since there is an infinite number of different input signals which could be applied to any system, it is usual to consider a subset of the most important, useful or common types of signals which the system is expected to encounter. The most common types of input signals used to evaluate performance in control engineering are impulse, step and ramp signals. Here, we focus on the response to step inputs, since this reveals the limitations of performance of a system when it is subject to rapidly changing inputs, e.g. the response of a receptor network to changes in ligand concentration.

For a first-order system with transfer function

$$G(s) = \frac{K}{\tau s + 1} \tag{2.35}$$

the step response has a simple exponential shape, as shown in Fig. 2.8, and the important measures of performance are the time constant τ, which corresponds to the time taken for the output to reach 63% of its final value, and the steady state value K.

Now consider the simple closed-loop control system shown in Fig. 2.9 with

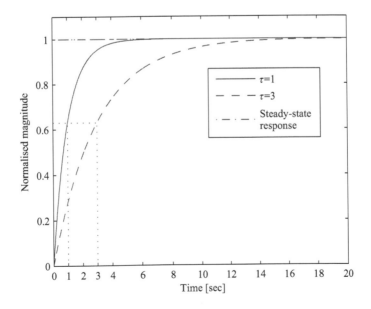

FIGURE 2.8: Step response of a first-order system for different values of the time constant.

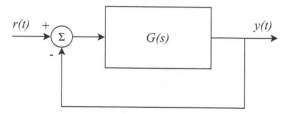

FIGURE 2.9: Closed-loop control system.

reference input $r(t)$, output $y(t)$ and

$$G(s) = \frac{K}{s(s+p)}.$$

The closed-loop transfer function for this system is given by

$$G(s) = \frac{Y(s)}{U(s)} = \frac{G(s)}{1+G(s)}$$

$$= \frac{K}{s^2+ps+K} \qquad (2.36)$$

and thus this is a second-order system. Adopting the generalised notation of Section 2.9, we can write the above equation as

$$Y(s) = \frac{\omega_n^2}{s^2+2\zeta\omega_n s+\omega_n^2}R(s) \qquad (2.37)$$

where $\omega_n = \sqrt{K}$ and $\zeta = \frac{p}{2\sqrt{K}}$. For a unit step input $R(s) = \frac{1}{s}$, we thus have

$$Y(s) = \frac{\omega_n^2}{s(s^2+2\zeta\omega_n s+\omega_n^2)} \qquad (2.38)$$

Inverse Laplace transforming gives the output $y(t)$ in the time domain as

$$y(t) = 1 - \frac{1}{\beta}e^{-\zeta\omega_n t}sin(\omega_n \beta t+\theta) \qquad (2.39)$$

The step response of this second-order system is shown in Fig. 2.10. On the figure are shown three of the most important time domain performance measures for feedback control systems — the rise time, the overshoot and the settling time. Note that the overall speed of response of the system is determined by the natural frequency ω_n, since this also determines the bandwidth of the system. However, the tradeoff between the *initial* speed of response of the system (as defined by the rise time) and the *accuracy* of the response (as defined by the overshoot and settling time) is encapsulated in the value

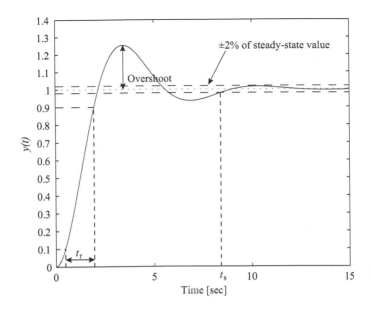

FIGURE 2.10: Characteristic parameters of the step response of a second-order system.

of a single parameter, the damping factor ζ. For $\zeta < 1$ the system is *under-damped*, and as the damping factor decreases the system exhibits a faster, but more oscillatory response, with larger initial overshoot of its target value and a longer settling time. For $\zeta > 1$, the system is termed *overdamped*, with no overshoot but with an increasingly sluggish response as ζ increases. A value of $\zeta = 1$ (termed *critical damping*) represents the optimal tradeoff between the conflicting objectives of speed and accuracy of the response (i.e. it gives the maximum speed of response for which no overshoot occurs). The step response of a second-order system for different values of the damping factor is shown in Fig. 2.11.

Note that the above limitations on performance hold exactly only in the case of second-order linear systems, and more complex systems incorporating nonlinear or adaptive behaviour may be able to get around them. However, they have been observed to hold at least approximately in very many types of systems traditionally studied in the field of control engineering, including at least one biological example which would appear on first glance to involve much more than second-order dynamics (consider the steering response of a car driver at different speeds — as the speed of response required of the driver increases, so does the overshoot and settling time, until eventually instability may be encountered!). Whether these tradeoffs are as widely conserved in biological systems as they appear to be in physical ones is an interesting open

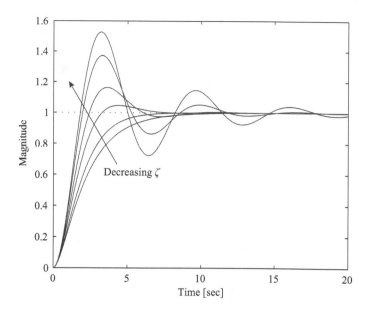

FIGURE 2.11: Step response of a second-order system for different values of the damping factor.

question that is just starting to be elucidated by current systems biology research. In any case, the performance measures introduced above are entirely generic, and may be applied to evaluate and compare the response of any type of dynamical system.

2.9 Characterising system dynamics in the frequency domain

In this section, we introduce some of the most important measures for characterising frequency domain dynamics, since these measures can provide valuable insight into the dominant processes underlying the response of biological systems to stimuli over different time scales. Following standard practice in control engineering, we focus on second-order systems, which may be considered a reliable idealisation of many different types of systems with more complex dynamics. In control engineering, a general second-order system with input u, output y and differential equation given by

$$a\ddot{y} + b\dot{y} + cy = Ku \qquad (2.40)$$

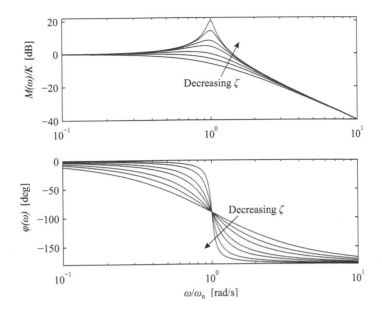

FIGURE 2.12: Bode plots of the normalised gain and the phase of the frequency response as a function of dimensionless frequency.

is typically written in the following standard form:

$$\ddot{y} + 2\zeta\omega_n\dot{y} + \omega_n^2 y = \omega_n^2 K u \tag{2.41}$$

with transfer function

$$G(s) = \frac{\omega_n^2 K}{s^2 + 2\zeta\omega_n s + \omega_n^2} \tag{2.42}$$

In the above expressions, ω_n is called the *natural frequency* of the system, and ζ is called the damping factor. Putting $s = j\omega$, we can compute the frequency response of this system as

$$G(j\omega) = \frac{\omega_n^2 K}{\omega_n^2 - \omega^2 + j2\zeta\omega_n\omega}$$

$$= \frac{K}{1 - (\omega/\omega_n)^2 + j2\zeta(\omega/\omega_n)} \tag{2.43}$$

Bode plots of the normalised gain and the phase of the frequency response as a function of dimensionless frequency ω/ω_n for different values of the damping factor ζ are shown in Fig. 2.12. From this figure the key characteristics of the frequency response are immediately apparent:

- The magnitude of an input signal of a given frequency will either be *amplified* or *attenuated* by the system, depending on whether the magnitude of $G(j\omega)$ at that frequency is greater than or less than 0 dB on the Bode magnitude plot. Harmonics of the input signal with frequencies close to the natural frequency will be most strongly amplified.

- The *steady state response* of the system, i.e. its response to constant input signals, is given by the values of the magnitude and phase on the extreme left of the Bode plot.

- The *roll-off rate* denotes the rate at which the attenuation of input signals increases with increasing frequency, and is calculated by determining the change in the Bode magnitude over one decade of frequency, e.g. -20 dB/decade.

- *phase lag* denotes the time lag between the input signal and the response of the system, and can be read from the Bode phase plot. From Fig. 2.12 we can see that for smaller values of the damping factor, the phase lag approaches 180° at lower and lower frequencies. The combination of large phase lags and strong signal amplification in lightly damped systems can easily lead to instability when these systems are subject to closed-loop feedback control.

- The *bandwidth* of the system denotes the range of input signal frequencies that can produce a significant response from the system. In control engineering, this is usually taken as being the range of frequencies where the frequency response lies within 3 dB of the peak magnitude.

2.10 Block diagram representations of interconnected systems

Another advantage of the type of frequency domain models described above is that they greatly simplify the computation of models of interconnected systems through the use of block diagram algebra. Although such interconnection schemes could in principle exhibit any topology, in practice we can identify some basic motifs, which will be presented next.

In a block diagram scheme, a variable (usually assumed to be scalar) is represented as a line ending in an arrow and a system is represented by a box with the transfer function written inside it; other basic elements are the sum node and the branch point. A common problem is, given a certain block diagram scheme, to calculate the transfer function between two signals in the scheme. In order to do this, we start with the basic interconnection schemes, namely the series, parallel and feedback schemes.

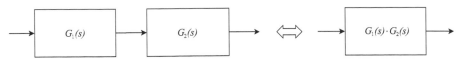

FIGURE 2.13: Series connection of two systems.

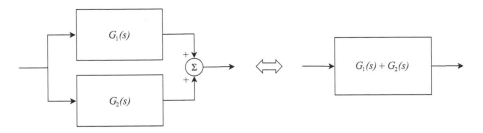

FIGURE 2.14: Parallel connection of two systems.

Series connection.
Two systems G_1 and G_2 are connected in series (or in cascade) when the output of the first system coincides with the input of the second one (see Fig. 2.13), so that

$$Y_2(s) = G_2(s)Y_1(s) = G_2(s)G_1(s)U_1(s).$$

Thus, if we take as input U_1 and as output Y_2, the transfer function will be $G(s) = G_2(s)G_1(s)$.

Parallel connection.
Two systems are connected in parallel (Fig. 2.14) if they have the same input and the total output is the sum of their outputs. In this case, the transfer function is

$$Y(s) = Y_1(s) + Y_2(s) = G_1(s)U_1(s) + G_2(s)U_2(s) = (G_1(s) + G_2(s))\,U(s).$$

Thus, the transfer function of the equivalent system is $G(s) = G_1(s) + G_2(s)$.

Feedback connection.
The standard feedback connection is shown in Fig. 2.15 and the scheme is also referred to as a closed-loop system. The transfer function can be easily derived from the relation

$$Y(s) = Y_1(s) = G_1(s)\,(U(s) - Y_2(s)) = G_1(s)\,(U(s) - G_2(s)Y_1(s)),$$

which yields

$$Y(s) = \frac{G_1(s)}{(1 + G_1(s)G_2(s))}U(s).$$

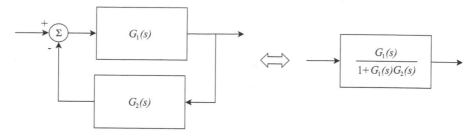

FIGURE 2.15: Negative feedback connection of two systems.

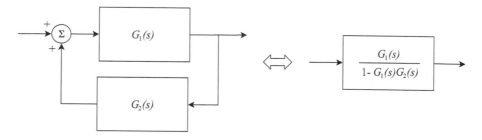

FIGURE 2.16: Positive feedback connection of two systems.

Note that the scheme in Fig. 2.15 is a negative feedback loop, since the output of the second system is subtracted from the total input. Negative feedback is typically used in control systems when regulating some value to a desired level. As we shall see in future chapters, other systems, which may be designed to generate sustained oscillations or exhibit switch-like behaviour, are often based on positive feedback (Fig. 2.16), in which case the transfer function reads

$$Y(s) = \frac{G_1(s)}{(1 - G_1(s)G_2(s))}U(s).$$

The next example shows how to exploit the basic rules illustrated above to derive the transfer function between two signals in a more involved block diagram.

Example 2.6
Given the block diagram in Fig. 2.17a, we want to derive the transfer function of the represented system. Note that in this block diagram it is not possible to distinguish series, parallel or feedback interconnections that can be isolated and reduced by applying the rules given above. Therefore, we first perform a little manipulation, by moving G_2 upstream of the first sum node, to derive the equivalent block diagram reported in Fig. 2.17b. Then, noting that the order of the sum nodes can be inverted and using the parallel and series

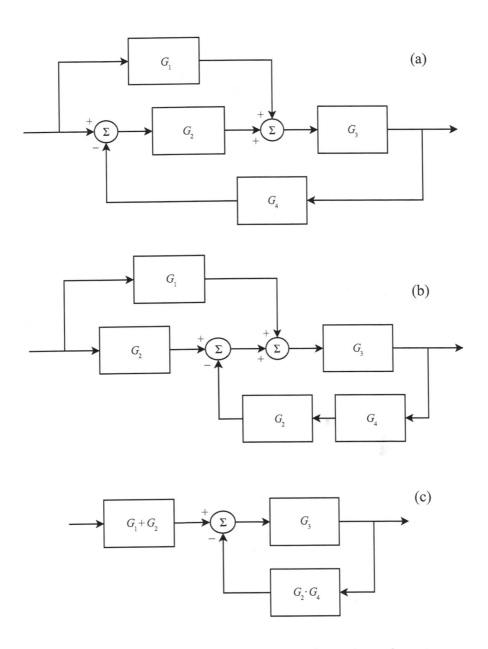

FIGURE 2.17: Simplification of a block diagram by applying the series, parallel and feedback interconnection rules.

interconnection rules, we find the block diagram shown in Fig. 2.17c. Finally, using the feedback and series interconnection rules, we find the input–output transfer function of the system, that is

$$G(s) = \frac{G_1(s)G_3(s) + G_2(s)G_3(s)}{1 + G_2(s)G_3(s)G_4(s)} .$$

⬜

2.11 Case Study I: Characterising the frequency dependence of osmo-adaptation in *Saccharomyces cerevisiae*

Biology background: Osmosis is the diffusion of water through a semi-permeable membrane (permeable to the solvent, but not the solute), from the compartment containing a low concentration (*hypotonic*) solution to the one at high concentration (*hypertonic*). The movement of fluid, while decreasing the concentration difference, increases the inner pressure of the hypertonic compartment, thus producing a force that counteracts osmosis; when these two effects balance each other, the osmotic equilibrium is reached. Osmosis is particularly important for cells, since many biological membranes are permeable to small molecules like water, but impermeable to larger molecules and ions. Osmosis provides the primary means by which water is transported into and out of cells; moreover the turgor pressure of a cell (i.e. the force exerted outward on a cell wall by the water contained in the cell) is largely maintained by osmosis between the cell interior and the surrounding hypotonic environment.

Osmotic shocks arise due to a sudden rise or fall in the concentration of a solute in the cell's environment, resulting in rapid movements of water through the cell's membrane. These movements can produce dramatic consequences for the cell — loss of water inhibits the transport of substrates and cofactors into the cell, while the uptake of large quantities of water can lead to swelling, rupture of the cell membrane or apoptosis. Due to their more direct contact with their environment, single-celled organisms are generally more vulnerable to osmotic shock. However, cells in large animals such as mammals also suffer similar stresses under certain conditions, [2].

Osmo-adaptation is the mechanism by which cells cope with large changes in the concentration of solutes in the environment, to avoid the aforementioned harmful consequences. Organisms have evolved a variety of mechanisms to respond to osmotic shock. To properly control gene expression, the cell must be able to sense osmotic changes and transmit an appropriate response signal to the nucleus. Typically, cells use surface sensors to gather information about the osmolarity of their surroundings; these sensors generate signals which activate signal transduction networks to coordinate the response of the cell, [3]. Recent experimental research indicates that most eukaryotic cells use the mitogen-activated protein 1 (MAP1) kinase pathways for this purpose, [4].

2.11.1   Introduction

Building an exhaustive mechanistic mathematical model of the osmotic shock
response would currently entail a significant research effort, notwithstanding
the fact that at present this goal is hindered by our incomplete knowledge
of the reactions and parameters involved. Even though the system dynamics
emerge from an intricate network of interactions, their main features can be
often ascribed to a limited number of important molecular regulatory mechanisms. A feasible approach to derive a concise description of these basic
mechanisms is to analyse the dynamics of the system in the frequency domain, especially since, in this case, the various subprocesses act at very different time-scales, e.g. ligand binding/unbinding, phosphorylation, diffusion between compartments and transcription of genes. As we shall see, the slow subprocesses predominantly dictate the dynamic evolution of the system, while
the fast ones can be assumed to be constantly at equilibrium (this assumption
is often referred to as *Quasi Steady State Approximation* in biochemistry).

In this Case Study, we apply the frequency domain analysis concepts introduced in the previous sections, in order to develop a concise model of the
high-osmolarity glycerol (HOG) mitogen-activated protein kinase (MAPK)
cascade in the budding yeast *Saccharomyces cerevisiae*. Our treatment is
based on the results presented in [5].

2.11.2   Frequency domain analysis

In *S. cerevisiae*, after a hyperosmotic shock, membrane proteins trigger a
signal transduction cascade that culminates in the activation of the MAPK
Hog1. This protein, which is normally cytoplasmic, is then imported into the
nucleus, where it activates several transcriptional responses to osmotic stress.
Hog1 is deactivated (through dephosphorylation) when the osmotic balance
is restored, thus allowing its export back to the cytoplasm.

As a first step towards model construction, we must determine the input(s)
and output(s) of the system we want to analyse: in this case the input is chosen
to be the extracellular osmolyte concentration and the output is the concentration of active (phosphorylated) Hog1. In the experiments presented in [5],
the input is manipulated by varying the salt concentration of the medium
surrounding the cells, whereas the output is measured by estimating the lo-
calisation of Hog1 in the nucleus through fluorescence image analysis. Thus,
the relative activity of Hog1 is measured as the nuclear to total Hog1 ratio
in the cell, $R(t)$, averaged over the 50–300 cells observed in the microscope's
field of view.

The cells have been shocked by supplying a medium with pulse wave (as in
the dashed signal in Fig. 2.4) changes in concentration of period T_0, alternating
the concentration level between 0.2 and 0 M NaCl. The experiments, repeated
with different values of T_0, ranging from 2 to 128 min, show that the steady
state response is approximately sinusoidal, with period T_0. Recall that, as

shown in Example 2.4, a sinusoidal input can be approximated reasonably well by a pulse–wave, which is much easier to reproduce experimentally: using a first harmonic approximation, the experimental input can therefore be written

$$u(t) \approx 0.2 \left(\frac{1}{2} + \frac{2}{\pi} \sin(\omega_0 t) \right) .$$

This, on the basis of the arguments illustrated in Section 2.3, suggests that the system behaviour can be approximately described by means of a linear model, at least over the time scale reported above. Therefore, we can assume that the steady-state response $R_\infty(t)$ takes the form

$$R_\infty(t) = M'(\omega_0) \sin (\omega_0 t + \varphi(\omega_0)) + R_0 , \qquad (2.44)$$

where $M'(\omega) = M(\omega) \, 0.4/\pi$, and R_0 is an offset term. The values $M'(\omega_0)$ and $\varphi(\omega_0)$ for different values of ω_0 can be computed by fitting the parameters in Eq. (2.44) to the experimental time response, as shown in Fig. 2.18 for $\omega_0 = 2\pi/8$ rad/min. The resulting sampled frequency response is shown on

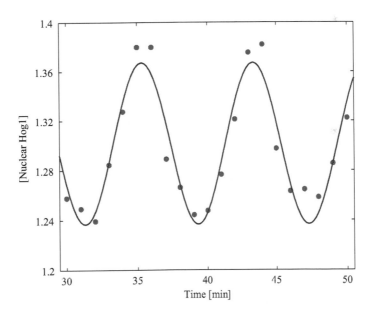

FIGURE 2.18: The function $R_\infty(t)$ in Eq. (2.44) (solid line), fitted to the experimental measurements of nuclear Hog1 enrichment (circles), obtained with a pulse wave of period $T_0 = 8$ min.

the Bode plots in Fig. 2.19. In order to gain further insight into the operation

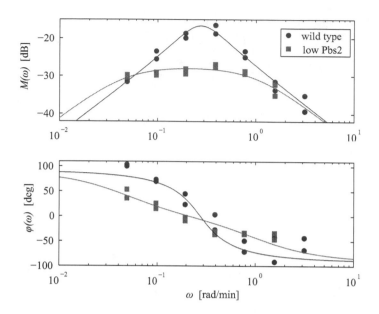

FIGURE 2.19: Frequency response of the osmo-adaptation system: experimental data (two measurements at each frequency) with wild type (circles) and underexpressed Pbs2 mutant (squares) strains. For each cell type, the figure shows the frequency response of system Eq. (2.45) with the parameters optimised through fitting against experimental data.

of the underlying regulatory mechanisms, we now proceed to develop an LTI model of the process: to this end, we can define a parametric transfer function $G(s)$ with n zeros and n poles (taking the form of Eq. (2.17)) and try to fit the associated frequency response $G(j\omega)$ to the experimental points for different values of n. Through this procedure we see that a satisfactory interpolation can be obtained with a second–order system, exhibiting a zero in the origin and a pair of complex conjugated poles

$$G(s) = K \frac{\omega_n^2 s}{s^2 + 2\zeta\omega_n s + \omega_n^2} \, . \tag{2.45}$$

The parameters K, ζ, ω_n can be computed by fitting the frequency response $G(j\omega)$ to the experimental points, as shown in Fig. 2.19. The best-fit parameters are shown in Table 2.2.

2.11.3 Time domain analysis

In order to assess the quality of the models, they have been used to predict the response of the two yeast strains to a step change in the input concen-

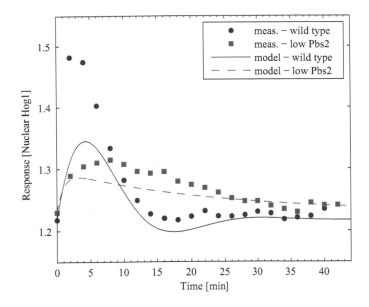

FIGURE 2.20: Time domain response of the osmo-adaptation system to a step change in concentration of amplitude 0.2 M NaCl: comparison of the responses predicted by the linear models developed in the frequency domain vs the experimental measurements.

tration of amplitude 0.2 M NaCl. The predicted responses of the models are compared with the experimental measurements in Fig. 2.20: the responses of the linear systems are offset by a constant value (1.23 M NaCl), which is the experimentally measured basal activity level of Hog1. The two models show a good qualitative match to the different sets of data for the two yeast strains (of course we do not expect a perfect match, since these are linear models of a process that will clearly also involve some nonlinear dynamics). Note that the wild type model exhibits a pair of complex conjugated poles ($\zeta < 1$), and therefore the response is oscillatory, with a larger overshoot and a faster response than the low Pbs2 model, as expected from the experimental data.

TABLE 2.2
Best-fitting parameters for model (2.45)

	K	ω_n	ζ
wild type	4.238	0.2787	0.5144
low Pbs2	6.978	0.2131	2.398

Indeed, the latter has two real poles ($\zeta > 1$), and thus exhibits a limited initial overshoot, a fast initial rise (due to the pole with small time constant) and a slow decay (caused by the large time constant associated with the other real pole).

The identified model can be translated into the state-space canonical form

$$\begin{pmatrix} \dot{x}_1 \\ \dot{x}_2 \end{pmatrix} = \begin{pmatrix} 0 & -\omega_n^2 \\ 1 & -2\zeta\omega_n \end{pmatrix} \begin{pmatrix} x_1 \\ x_2 \end{pmatrix} + \begin{pmatrix} 0 \\ K\omega_n^2 \end{pmatrix} u \qquad (2.46a)$$

$$y = x_2 \qquad (2.46b)$$

Since the second state variable coincides with the observable output of the system, it can be readily associated with a physical quantity of the process (the level of Hog1 activity), and thus it is convenient to leave it unchanged. However, the *hidden* state variable x_1 can be arbitrarily substituted with a new one, denoted by x_1', using the linear transformation

$$\begin{pmatrix} x_1 \\ x_2 \end{pmatrix} = \begin{pmatrix} \alpha & \beta \\ 0 & 1 \end{pmatrix} \begin{pmatrix} x_1' \\ x_2 \end{pmatrix}$$

which is parameterised with respect to α and β. Letting $\alpha = -1$, we obtain the new state-space representation

$$\begin{pmatrix} \dot{x}_1' \\ \dot{x}_2 \end{pmatrix} = \begin{pmatrix} -\beta & \omega_n^2 + \beta^2 - 2\beta\zeta\omega_n \\ -1 & \beta - 2\zeta\omega_n \end{pmatrix} \begin{pmatrix} x_1' \\ x_2 \end{pmatrix} + \begin{pmatrix} \beta K\omega_n^2 \\ K\omega_n^2 \end{pmatrix} u \qquad (2.47a)$$

$$y = x_2 \qquad (2.47b)$$

which corresponds to the block diagram in Fig. 2.21. The transformation has been chosen such that the hidden variable is directly compared with the (scaled) input in the block diagram of the system (Fig. 2.21). This enables us to assign a physical meaning also to this quantity: since the input is the external pressure (or, equivalently, the osmolyte concentration in the external environment), x_1 represents the internal pressure (or, equivalently, the cellular osmolyte concentration). This representation provides some interesting insights into the inner mechanisms of the hyperosmotic shock response: the model structure tells us that the response is partly mediated by the Hog1 MAPK pathway and partly by a second pathway, which is independent of Hog1. Since Hog1 is activated by Pbs2, we can derive useful insights by comparing the responses of the wild type strain with the mutant strain (in which pbs2 is underexpressed). This comparison suggests that the feedback action provided by the Hog1 pathway is stronger, producing a faster response.

The hyperosmotic shock response has been thoroughly studied in the biological literature; therefore, it is also interesting to see if the quantitative description derived above is in agreement with the available qualitative models and experimental results. The biological literature states that the regulation of the hyperosmotic shock response in *S. cerevisiae* is actually implemented through two distinct mechanisms:

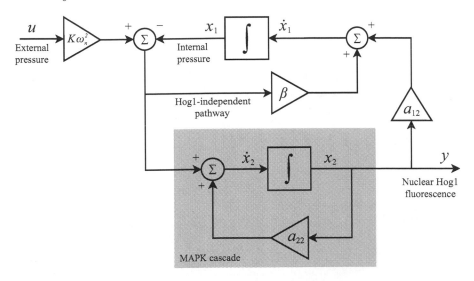

FIGURE 2.21: Block diagram representation of system (2.47). The values of the gain blocks a_{12} and a_{22} are equal to the corresponding entries of the A matrix of the state-space representation (2.47).

a) The activity of the membrane protein Fps1 is regulated so as to decrease the glycerol export rate; this mechanism is Hog1 independent and is activated in less than 2 minutes.

b) A second pathway, dependent on the activation of Hog1, increases the expression of Gpd1 and Gpp2 which, in turn, accelerate the production of glycerol; this response is known to be significantly slower than the previous one, causing an increase in the intracellular glycerol concentration after ~ 30 minutes.

Looking at both the model predictions and the experimental data, we note that the peak times of the responses of both the wild type and mutant strains are less than ten minutes. Thus, in both cases the response is much faster than the characteristic dynamics of gene expression involved in the regulatory mechanism b). Therefore, the difference in the responses of the two strains has to be ascribed to changes at the protein–protein interaction level. This suggests that the MAPK Hog1 plays a role not only in the transcriptional regulation of glycerol producing proteins, but also in the control of rapid glycerol export, as supported also by experimental studies [6].

2.12   Case Study II: Characterising the dynamics of the *Dictyostelium* external signal receptor network

Biology background: *Dictyostelium discoideum* are social amoebae which live in forest soil and feed on bacteria such as *Escherichia coli* that are found in the soil and in decaying organic matter. *Dictyostelium* cells grow independently, but under conditions of starvation they initiate a well-defined program of development [7]. In this program, the individual cells aggregate by sensing and moving towards gradients in cAMP (cyclic Adenosine Mono-Phosphate), a process known as chemotaxis. As they move along the cAMP gradient, the individual cells bump into each other and stick together through the use of glycoprotein adhesion molecules, to form complexes of up to 10^5 cells.

Subsequently, the individual cells form a slime mold which eventually becomes a fruiting body which emits spores. The early stage of aggregation in *Dictyostelium* cells is initiated by the production of spontaneous oscillations in the concentration of cAMP (and several other molecular species) inside the cell. The oscillations in each individual cell are not completely autonomous, but are excited by changes in the concentration of external cAMP, which is secreted from each cell and diffused throughout the region where the cells are distributed. Many of the processes employed by *Dictyostelium* to chemotax are shared with other organisms, including mammalian cells.

Chemotaxis occurs to some extent in almost every cell type at some time during its development and it is a major component of the inflammatory and wound-healing responses, the development of the nervous system as well as tumour metastasis. Chemotaxis and signal transduction by chemoattractant receptors play a key role in inflammation, arthritis, asthma, lymphocyte trafficking and also in axon guidance. For this reason, *Dictyostelium* is a very useful model organism for the study of signal transduction mechanisms implicated in human disease.

Recent examples of *Dictyostelium*-based biomedical research include the analysis of immune cell disease and chemotaxis, centrosomal abnormalities and lissencephaly, bacterial intracellular pathogenesis and mechanisms of neuroprotective and anti-cancer drug action, [8]. Other advantages of *Dictyostelium* as a model organism include the fact that they can be easily observed at organismic, cellular and molecular levels, primarily because of their restricted number of cell types, behaviours and their rapid growth. The entire genome of *Dictyostelium* has been sequenced and is available online in a model organism database called dictyBase.

2.12.1 Introduction

In cellular signal transduction, external signalling molecules, called ligands, are initially bound by receptors which are distributed on the cell surface. The ligand–receptor complex then initiates various signal transduction pathways, such as activation of immune responses, growth factors, etc. Inappropriate activation of signal transduction pathways is considered to be an important factor underlying the development of many diseases. Hence, robust performance of ligand and receptor interaction networks constitutes one of the crucial mechanisms for ensuring the healthy development of living organisms. In a recent study, [9], a generic model structure for ligand–receptor interaction networks was proposed. Analysis of this model showed that the ability to capture ligand together with the ability to internalise bound-ligand complexes are the key properties distinguishing the various functional differences in ligand–receptor interaction networks. From the perspective of control engineering, it is also tempting to speculate that nature will have evolved the dynamic behaviour in such structural networks to deliver robust and optimal performance in relaying external signals into the cell, [10, 11].

In this Case Study, we show that the ligand–receptor interaction network employed to relay external cAMP signals in aggregating *Dictyostelium discoideum* cells appears to exhibit such generic structural characteristics. We also use both frequency and time domain analysis techniques of the kind described earlier in this chapter to investigate the underlying control mechanisms for this system. We show that the network parameters for the ligand-bound cell receptors which are distributed on the outer shell of *Dictyostelium discoideum* cells are highly optimised, in the sense that the response speed is the fastest possible while ensuring that no overshoot occurs for step changes in external signals. We also show that the bandwidth of the network is just above the minimum necessary to deliver adequate tracking of the type of oscillations in cAMP which have been observed experimentally in *Dictyostelium* cells during chemotaxis. Our treatment follows that of [12].

2.12.2 A generic structure for ligand–receptor interaction networks

In [9], a generic structure was proposed for cellular ligand–receptor interaction networks of the following form:

$$L + R \underset{k_{\text{off}}}{\overset{k_{\text{on}}}{\rightleftharpoons}} C, \ Q_R \to R, \ f(t) \to L, \tag{2.48a}$$

$$R \xrightarrow{k_t} \emptyset, \ C \xrightarrow{k_e} \emptyset \tag{2.48b}$$

We denote by L the concentration of ligand, R is the number of external cell receptor molecules, C is the number of ligand-receptor complex molecules, k_{on} is the forward reaction rate for ligands binding to receptors, k_{off} is the

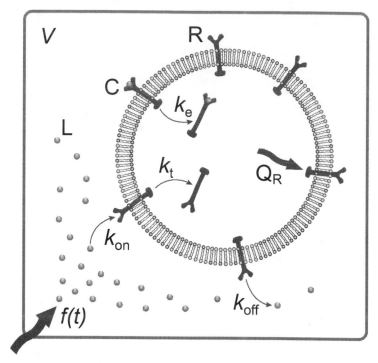

FIGURE 2.22: A generic model for ligand–receptor interactions, [9].

reverse reaction rate for ligands dissociating from receptors, k_t is the rate of internalisation of receptor molecules and k_e is the rate of internalisation of ligand-receptor complexes. Q_R is equal to $R_T \times k_t$, where R_T is the steady-state number of cell surface receptors when $C = 0$ and $L = 0$, $\emptyset$ is the null species of either the receptor or the complex, $f(t)$ is some input signal and t is time — see Fig. 2.22. The corresponding differential equations are given by

$$\frac{d}{dt}\begin{bmatrix} R \\ C \\ L \end{bmatrix} = \begin{bmatrix} -k_{on}RL + k_{off}C - k_tR + Q_R \\ k_{on}RL - k_{off}C - k_eC \\ (-k_{on}RL + k_{off}C)/(N_{av}V_c) + f(t) \end{bmatrix} \qquad (2.49)$$

where N_{av} is Avogadro's number, 6.023×10^{23}, and V_c is the cell volume in liters. In normalised form, the above equation can be written as

$$\frac{d}{dt^*}\begin{bmatrix} R^* \\ C^* \\ L^* \end{bmatrix} = \begin{bmatrix} -R^*L^* + C^* - \alpha(R^* - 1) \\ R^*L^* - C^* - \beta C^* \\ \gamma(-R^*L^* + C^*) + u \end{bmatrix} \qquad (2.50)$$

where $t^* = k_{off}\, t$, $R^* = R/R_T$, $C^* = C/R_T$, $L^* = L/K_D$, $u = f(t)/k_{off}/K_D$ and K_D is the receptor dissociation constant, i.e., $K_D = k_{off}/k_{on}$. In the normalised model, α is a quantity proportional to the probability of internalisation of unbound receptors, β is a quantity proportional to the probability of internalisation of captured ligand by receptors before dissociation of the ligand from the receptors, and γ represents the level of sensitivity of the receptors to the external signals [9]. By assuming that the number of receptors is much larger than the number of ligands, i.e. $dR/dt \approx 0$ $(R \approx R_T)$, the following simplified ligand and ligand–complex kinetics are obtained:

$$\frac{d}{dt^*} \begin{bmatrix} C^* \\ L^* \end{bmatrix} = \begin{bmatrix} -(1+\beta) & 1 \\ \gamma & -\gamma \end{bmatrix} \begin{bmatrix} C^* \\ L^* \end{bmatrix} + \begin{bmatrix} 0 \\ 1 \end{bmatrix} u \qquad (2.51)$$

where β and γ are given by

$$\beta = \frac{k_e}{k_{off}}, \quad \gamma = \frac{K_a R_T}{N_{av} V_c}, \qquad (2.52)$$

and $K_a = 1/K_D = k_{on}/k_{off}$ is the association constant.

2.12.3 Structure of the ligand–receptor interaction network in aggregating *Dictyostelium* cells

We now show how a ligand–receptor interaction network displaying the generic structure given above may be extracted in a straightforward manner from a model for the complete network underlying cAMP oscillations in *Dictyostelium* published in [7, 13], and shown in Fig. 2.23. In this network, cAMP is produced inside the cell when adenylyl cyclase (ACA) is activated after the binding of extracellular cAMP to the surface receptor CAR1. Ligand-bound CAR1 activates the mitogen activated protein kinase (ERK2) which in turn inhibits the cAMP phosphodiesterase RegA by phosphorylating it. When cAMP accumulates internally, it activates the protein kinase PKA by binding to the regulatory subunit of PKA. ERK2 is inactivated by PKA and hence can no longer inhibit RegA by phosphorylating it. A protein phosphatase activates RegA such that RegA can hydrolyse internal cAMP. Either directly or indirectly, CAR1 is phosphorylated when PKA is activated, leading to loss-of-ligand binding. When the internal cAMP is hydrolysed by RegA, PKA activity is inhibited by its regulatory subunit, and protein phosphatase(s) returns CAR1 to its high-affinity state. Secreted cAMP diffuses between cells before being degraded by the secreted phosphodiesterase PDE. For more details of the experimental results upon which the various interactions in the above network are based, the reader is referred to [7].

The dynamics of the network shown in Fig. 2.23 can be expressed as a set of nonlinear differential equations with kinetic constants k_{1-14}. The activity of each of the seven components in the network is determined by the balance between activating and inactivating enzymes which is then reflected in the equations in the form of an activating and deactivating term. The model thus

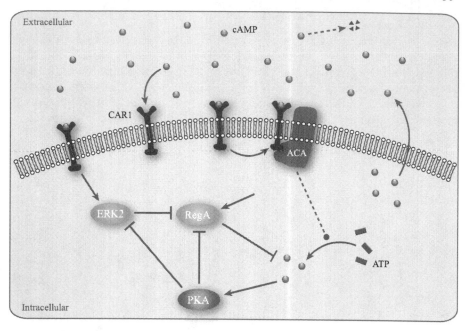

FIGURE 2.23: The model of [7] for the network underlying cAMP oscillations in *Dictyostelium*. The normal arrows and the broken arrows represent activation and self-degradation, respectively. The bar arrows represent inhibition.

consists of a set of nonlinear differential equations in the following form:

$$\frac{d\,ACA}{dt} = k_1 CAR1 - k_2 ACA\,PKA$$

$$\frac{d\,PKA}{dt} = k_3 cAMPi - k_4 PKA$$

$$\frac{d\,ERK2}{dt} = k_5 CAR1 - k_6 PKA\,ERK2$$

$$\frac{d\,RegA}{dt} = k_7 - k_8 ERK2\,RegA \qquad (2.53)$$

$$\frac{d\,cAMPi}{dt} = k_9 ACA - k_{10} RegA\,cAMPi$$

$$\frac{d\,cAMPe}{dt} = k_{11} ACA - k_{12} cAMPe$$

$$\frac{d\,CAR1}{dt} = k_{13} cAMPe - k_{14} CAR1$$

where *cAMPi* and *cAMPe* are internal and external cAMP, respectively. The ligand–receptor interaction network for this model can be extracted as follows:

$$\frac{d}{dt}\begin{bmatrix} CAR1 \\ cAMPe \end{bmatrix} = \begin{bmatrix} -k_{14} & k_{13} \\ 0 & -k_{12} \end{bmatrix}\begin{bmatrix} CAR1 \\ cAMPe \end{bmatrix} + \begin{bmatrix} 0 \\ k_{11} \end{bmatrix}ACA \qquad (2.54)$$

Note that in the above, $CAR1$, $cAMPe$ and ACA are concentrations in units of μM, and k_{11}, k_{12}, k_{13} and k_{14} are reaction constants in units of $1/\min$. To transform the unit of $CAR1$ concentration into the number of molecules, we use the relation $C = CAR1 \ N_{av}V_c$ and hence derive the following:

$$\frac{dC}{dt} = -k_{14} \ CAR1 \ N_{av}V_c + k_{13}cAMPe \ N_{av}V_c$$
$$= -k_{14}C + k_{13}N_{av}V_cL \qquad (2.55)$$

where $L = cAMPe$. In addition,

$$\frac{dL}{dt} = -k_{12}L + k_{11}ACA \qquad (2.56)$$

With the normalised states,

$$\frac{dC^*}{dt^*} = -\frac{k_{14}}{k_{\text{off}}}C^* + \frac{k_{13}N_{av}V_c}{R_Tk_{\text{on}}}L^* \qquad (2.57)$$

Then,

$$\frac{dC^*}{dt^*} = -\frac{k_{14}}{k_{\text{off}}}C^* + L^{**} \qquad (2.58)$$

where $L^{**} = L^*K_L$ and $K_L = (k_{13}N_{av}V_c)/(R_Tk_{\text{on}})$. Note that K_L is multiplied by L^* to make the coefficient equal to one as in Eq. (2.51). Similarly,

$$\frac{dL^{**}}{dt^*} = -\frac{k_{12}}{k_{\text{off}}}L^{**} + u \qquad (2.59)$$

This can be written in a compact form as:

$$\frac{d}{dt}\begin{bmatrix} C^* \\ L^{**} \end{bmatrix} = \begin{bmatrix} -\dfrac{k_{14}}{k_{\text{off}}} & 1 \\ 0 & -\dfrac{k_{13}}{k_{\text{off}}} \end{bmatrix}\begin{bmatrix} C^* \\ L^{**} \end{bmatrix} + \begin{bmatrix} 0 \\ 1 \end{bmatrix}u \qquad (2.60)$$

Comparing Eq. (2.60) with Eq. (2.51), we notice that the only difference in the structure of the two equations is due to the effect of the $k_{\text{off}}C$ term in Eq. (2.49). However, in the case of the *Dictyostelium* network, it is reasonable to assume that the magnitude of the $k_{\text{off}}C$ term in Eq. (2.49) is negligible compared to the other terms, i.e. the rate of dissociation of the ligand from the receptor is very low. This is because efficient operation of the positive feedback loop involving external cAMP is crucial in maintaining the stable oscillations

in cAMP that are required for aggregation of the individual *Dictyostelium*
cells. Under this assumption, Eq. (2.51) can be rewritten as follows:

$$\frac{d}{dt^*}\begin{bmatrix} C^* \\ L^{**} \end{bmatrix} = \begin{bmatrix} -\beta & 1 \\ 0 & -\gamma \end{bmatrix}\begin{bmatrix} C^* \\ L^{**} \end{bmatrix} + \begin{bmatrix} 0 \\ 1 \end{bmatrix}u \qquad (2.61)$$

with

$$\beta = \frac{k_{14}}{k_{\text{off}}}, \quad \gamma = \frac{k_{12}}{k_{\text{off}}}, \quad u = \frac{k_{11}K_L\,ACA}{K_D k_{\text{off}}} \qquad (2.62)$$

and thus we see that the *Dictyostelium* receptor network displays the same
generic ligand-receptor interaction structure proposed in [9].

The values of the constants in the above equations are given as follows:
$k_{11} = 0.7$ min^{-1}, $k_{12} = 4.9$ min^{-1}, $k_{13} = 23.0$ min^{-1}, $k_{14} = 4.5$ min^{-1},
$R_T = 4 \times 10^4$, [14, 13], $k_{\text{off}} = 0.7 \times 60$ min^{-1} and $k_{\text{on}} = 0.7 \times 60 \times 10^7$ M^{-1}
min^{-1} [15]. Hence, $\beta = 0.107$ and $\gamma = 0.117$. In [16], the average diameter
and volume of a *Dictyostelium* cell are given by 10.25 μm and 565 μm^3. To
calculate V_c, we consider an approximation for the shape of a *Dictyostelium*
cell as a cylinder, and calculate the effective volume such that the maximum
number of ligand-bound CAR1 molecules is about 1% of the total number of
receptors, to give a value of V_c equal to 1.66×10^{-16} liters, [12].

2.12.4   Dynamic response of the ligand–receptor interaction network in *Dictyostelium*

In this section we investigate the time and frequency domain performance
of the *Dictyostelium* ligand–receptor interaction network, using the analysis
techniques introduced earlier in this chapter. Differentiating both sides of
Eq. (2.58) with respect to the normalised time, t^*, we get

$$\begin{aligned}
\frac{d^2C^*}{dt^{*2}} &= \frac{-k_{14}}{k_{\text{off}}}\frac{dC^*}{dt^*} + \frac{dL^{**}}{dt^*} \\
&= \frac{-k_{14}}{k_{\text{off}}}\frac{dC^*}{dt^*} - \frac{k_{12}}{k_{\text{off}}}\left(\frac{dC^*}{dt^*} + \frac{k_{14}}{k_{\text{off}}}C^*\right) + u
\end{aligned} \qquad (2.63)$$

In a compact form, this can be written as

$$\ddot{C}^* + \frac{k_{12} + k_{14}}{k_{\text{off}}}\dot{C}^* + \frac{k_{12}k_{14}}{k_{\text{off}}^2}C^* = u \qquad (2.64)$$

where the single and the double dots represent $d(\cdot)/dt^*$ and $d^2(\cdot)/dt^{*2}$, respectively.

Since the above equation is simply a second-order linear ordinary differential
equation, we can define the natural frequency, ω_n, and the damping ratio, ζ,
in the standard way as follows:

$$\ddot{C}^* + 2\zeta\omega_n\dot{C}^* + \omega_n^2 C^* = u \qquad (2.65)$$

Comparing Eq. (2.64) with Eq. (2.65) we have that

$$\omega_n = \frac{\sqrt{k_{12}k_{14}}}{k_{\text{off}}}, \quad \zeta = \frac{k_{12} + k_{14}}{2\sqrt{k_{12}k_{14}}} \tag{2.66}$$

Substituting the appropriate values for the *Dictyostelium* network, we find that ω_n is equal to 0.112 and ζ is equal to 1.001. Note that the overshoot, M_p, and the settling time, t_s, for a step input are given by [17]

$$M_p = \begin{cases} e^{-\pi\zeta\sqrt{1-\zeta^2}}, & \text{for } 0 \leq \zeta < 1 \\ 0, & \text{for } \zeta \geq 1 \end{cases} \tag{2.67}$$

$$t_s = \frac{-\ln 0.01}{\zeta\omega_n} \tag{2.68}$$

Thus, the kinetics of the *Dictyostelium* ligand–receptor network produce a system with a damping ratio almost exactly equal to 1, i.e. the critical damping ratio. As noted earlier in this chapter, the critical damping ratio is the optimal solution for maximising the speed of a system's response without allowing any overshoot:

$$\zeta^* = \arg\min J(\zeta) = t_s \tag{2.69}$$

subject to $M_p = 0$ and Eq. (2.64). Thus, it appears that *Dictyostelium* cells may have evolved a receptor/ligand interaction network which provides an optimal trade-off between maximising the speed of response and prohibiting overshoot of the response to external signals. Using the generic structure for

TABLE 2.3
Kinetic parameters for EGFR, TfR and VtgR [9]

	k_e	k_{off}	K_a [1/M]	R_T	V_c
EGFR	0.15	0.24	$10^9/2.47$	2×10^5	4×10^{-10}
TfR	0.6	0.09	$10^9/29.8$	2.6×10^4	4×10^{-10}
VtgR	0.108	0.07	$10^9/1300$	2×10^{11}	4×10^{-10}

ligand–receptor interaction networks proposed in [9], the speed of response of the *Dictyostelium* ligand-receptor kinetics may be compared with that of some other ligand-receptor kinetics, such as the epidermal growth factor receptor (EGFR), the transferrin receptor (TfR) and the vitellogenin receptor (VtgR). These receptors are involved in the development of tumours, the uptake of iron and the production of egg cells, respectively; see [18, 19, 20] for details. Using the definitions in Eq. (2.52) and the values given in Table 2.3, the damping factors for EGFR, TfR and VtgR may be calculated as follows: $\zeta_{\text{EGFR}} = 2.14$, $\zeta_{\text{TfR}} = 24.68$ and $\zeta_{\text{VtgR}} = 10.21$. Thus, for each of the above ligand-receptor kinetics, the responses are overdamped and thus the possibility of overshoot is completely prohibited. Indeed, in the case of the *Dictyostelium* network, the response cannot be under-damped for any combination of the kinetic parameters. This can be seen by considering the fact that

$$\zeta = \frac{k_{12} + k_{14}}{2\sqrt{k_{12}k_{14}}} \geq 1 \Rightarrow (k_{12} + k_{14})^2 \geq 4k_{12}k_{14}$$

$$\Rightarrow k_{12}^2 - 2k_{12}k_{14} + k_{14}^2 \geq 0 \Rightarrow (k_{12} - k_{14})^2 \geq 0 \tag{2.70}$$

for all $k_{12} > 0$ and $k_{14} > 0$. Hence, the overdamped dynamical response appears to stem from the network structure itself, rather than being dependent on any particular values of the kinetic parameters. The step responses with k_{12} and k_{14} perturbed by up to $\pm 50\%$ are shown in Fig. 2.24.

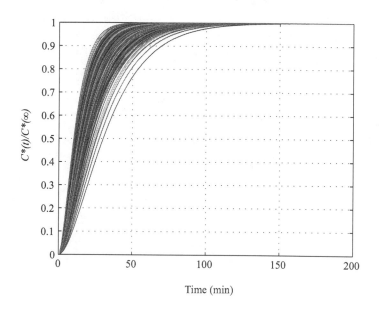

FIGURE 2.24: Step responses with the perturbed parameters k_{12} and k_{14}. Each kinetic parameter is perturbed by up to $\pm 50\%$. The response is normalised by the value of each steady state.

For this level of uncertainty in the kinetic parameters, the settling times vary between 35 min and 105 min (for the nominal parameter values the settling time is about 52 min).

One significant difference between the *Dictyostelium* network and the other ligand-receptor networks considered above is its relatively fast response time.

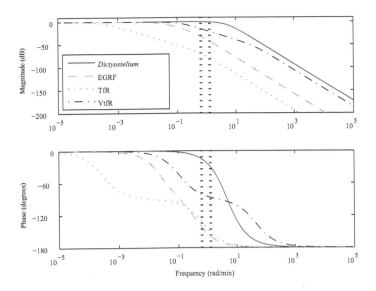

FIGURE 2.25: Bode plots for the *Dictyostelium*, EGFR, TfR and VtfR networks, where the magnitude is normalised by the magnitude at the lowest frequency for comparison. The region inside the two dashed vertical lines corresponds to oscillations with periods between 5 and 10 mins, which is the range of cAMP oscillations observed experimentally in the early stages of aggregation of *Dictyostelium*.

Since aggregating *Dictyostelium* cells exhibit oscillatory behaviour, rather than converging to a constant steady state, the ligand–receptor interaction network may have evolved to maximise the speed of response, in order to ensure the generation of robust and stable limit cycles in the concentration of cAMP. This can be more clearly seen in the Bode plots for the responses of the different networks, which are shown in Fig. 2.25. The bandwidth of the *Dictyostelium* ligand-receptor kinetics is about 3 rad/min, which is just above the minimum necessary to facilitate the oscillations in cAMP with a period of 5 to 10 min observed in *Dictyostelium* during chemotaxis.

References

[1] Dorf RC and Bishop RH. *Modern Control Systems* (Eighth edition). Boston: Addison-Wesley, 1998.

[2] Ho SN. Intracellular water homeostasis and the mammalian cellular osmotic stress response. *Journal of Cell Physiology*, 206(1):9–15, 2006.

[3] Rep M, Krantz M, Thevelein JM, and Hohmann S. The transcriptional response of *Saccharomyces cerevisiae* to osmotic shock. *The Journal of Biological Chemistry*, 275:8290–8300, 2000.

[4] Kultz D and Burg M. Evolution of osmotic stress signaling via MAP kinase cascades. *Journal of Experimental Biology*, 201(22):3015–3021, 1998.

[5] Mettetal JT, Muzzey D, Gómez-Uribe C, and van Oudenaarden A. The frequency dependence of osmo-adaptation in *Saccharomyces cerevisiae*. *Science*, 319(5862):482–484, 2008.

[6] Thorsen M et al. The MAPK Hog1p modulates Fps1p-dependent arsenite uptake and tolerance in yeast. *Molecular Biology of the Cell*, 17:4400–4410, 2006.

[7] Laub MT and Loomis WF. A molecular network that produces spontaneous oscillations in excitable cells of *Dictyostelium*. *Molecular Biology of the Cell*, 9:3521–3532, 1998.

[8] Williams RSB, Boeckeler K, Graf R, Muller-Taubenberger A, Li Z, Isberg RR, Wessels D, Soll DR, Alexander H, and Alexander S. Towards a molecular understanding of human diseases using *Dictyostelium discoideum*. *Trends in Molecular Medicine*, 12(9):415–424, 2006.

[9] Shankaran H, Resat H, and Wiley HS. Cell surface receptors for signal transduction and ligand transport: A design principles study. *PLoS Computational Biology*, 3(6):e101, 2007.

[10] Barkai N and Leibler S. Robustness in simple biochemical networks. *Nature*, 387(26):913–917, 1999.

[11] Csete ME and Doyle JC. Reverse engineering of biological complexity. *Science*, 295:1664–1669, 2002.

[12] Kim J, Heslop-Harrison P, Postlethwaite I, and Bates DG. Identification of optimality and robustness in *Dictyostelium* external signal receptors. In *Proceedings of the IFAC World Congress on Automatic Control*, Seoul, Korea, 2008.

[13] Maeda M, Lu S, Shaulsky G, Miyazaki Y, Kuwayama H, Tanaka Y, Kuspa A, and Loomis WF. Periodic signaling controlled by an oscillatory circuit that includes protein kinases ERK2 and PKA. *Science*, 304(5672):875–878, 2004.

[14] Bankir L, Ahloulay M, Devreotes PN, and Parent CA. Extracellular cAMP inhibits proximal reabsorption: Are plasma membrane cAMP

receptors involved? *American Journal of Physiology — Renal Physiology*, 282:376–392, 2002.

[15] Ishii D, Ishikawa KL, Fujita T, and Nakazawa M. Stochastic modelling for gradient sensing by chemotactic cells. *Science and Technology of Advanced Materials*, 5:715–718, 2004.

[16] Soll DR, Yarger J, and Mirick M. Stationary phase and the cell cycle of *Dictyostelium discoideum* in liquid nutrient medium. *Journal of Cell Science*, 20(3):513–523, 1976.

[17] Franklin GF, Powell JD, and Emani-Naeini A. *Feedback Control of Dynamic Systems*. Boston: Addison-Wesley, 3rd edition, 1994.

[18] Jorissen RN, Walker F, Pouliot N, Garrett TPJ, Ward CW, and Burgess AW. Epidermal growth factor receptor: mechanisms of activation and signalling. *Experimental Cell Research*, 284:31–53, 2003.

[19] Rao K, Harford JB, Rouault T, McClelland A, Ruddle FH, and Klausner RD. Transcriptional regulation by iron of the gene for the transferrin receptor. *Molecular and Cell Biology*, 6(1):236–240, 1986.

[20] Li A, Sadasivam A, and Ding JL. Receptor-ligand interaction between vitellogenin receptor (VtgR) and vitellogenin (Vtg), implications on low density lipoprotein receptor and apolipoprotein B/E. *Journal of Biological Chemistry*, 278(5):2799–2806, 2003.

3

Nonlinear systems

3.1 Introduction

Nonlinearity appears to be a fundamental property of biological systems. One reason for this may be the inherent complexity of biology — in the physical world, linear equations such as Newton's, Maxwell's and Schroedinger's are immensely successful descriptions of reality, but they are essentially equations of forces in a vacuum. Nonlinearity is fundamental in generating qualitative structural changes in complex phenomenon such as the transition from laminar to turbulent flow, or in phase changes from gas to liquid to solid. Whenever there are phase changes, whenever structure arises, nonlinear dynamics are often responsible, and the very fact that biological phenomena have for many years been successfully described in qualitative terms indicates the importance of nonlinearity in biological systems. As argued in [1], if it were not for nonlinearity, we would all be quivering jellies!

More concretely, even the briefest consideration of the dynamics which arise from the biochemical reaction kinetics underpinning almost all cellular processes, [2], reveals the ubiquity of nonlinear phenomena. The fundamental law of mass action states that when two molecules A and B react upon collision with each other to form a product C

$$A + B \xrightarrow{k} C \tag{3.1}$$

the rate of the reaction is proportional to the number of collisions per unit time between the two reactants and the probability that the collision occurs with sufficient energy to overcome the free energy of activation of the reaction. Clearly, the corresponding differential equation

$$\frac{dC}{dt} = kAB \tag{3.2}$$

where k is the temperature dependent reaction rate constant, is nonlinear. In enzymatic reactions, proteins called enzymes catalyse (i.e. increase the rate of) the reaction by lowering the free energy of activation. This situation may be represented by the Michaelis–Menten model, which describes a two-step process whereby an enzyme E first combines with a substrate S to form a complex C, which then releases E to form the product P

$$S + E \underset{k_2}{\overset{k_1}{\rightleftharpoons}} C \overset{k_3}{\rightarrow} P + E \tag{3.3}$$

The corresponding differential equation relating the rate of formation of the product to the concentrations of available substrate and enzyme is again nonlinear

$$\frac{dP}{dt} = \frac{V_{max}S}{K_m + S} \tag{3.4}$$

where the equilibrium constant $K_m = (k_2 + k_3)/k_1$ and the maximum reaction velocity $V_{max} = k_3 E$. Cooperativity effects, where the binding of one substrate molecule to the enzyme affects the binding of subsequent ones, serve to further increase the nonlinearity of the underlying kinetics. In general, for n substrate molecules with n equilibrium constants K_{m1} through K_{mn}, the rate of reaction is given by the Hill equation

$$\frac{dP}{dt} = \frac{V_{max}S^n}{K_h^n + S^n} \tag{3.5}$$

where $K_h^n = \prod_{i=1}^{n} K_{mn}$.

Nonlinear Michaelis–Menten and Hill-type functions are also ubiquitous in higher-level models of cellular signal transduction pathways and transcriptional regulation networks. In transcriptional regulatory networks, for example, transcription and translation may be considered as dynamical processes, in which the production of mRNAs depends on the concentrations of protein transcription factors (TFs) and the production of proteins depends on the concentrations of mRNAs. Equations describing the dynamics of transcription and translation, [3], can then be written as

$$\frac{dm_i}{dt} = g_i(\mathbf{p}) - k_i^g m_i$$

$$\frac{dp_i}{dt} = k_i m_i - k_i^p p_i$$

respectively, where m_i and p_i denote mRNA and protein concentrations and k_i^g and k_i^p are the degradation rates of mRNA i and protein i. The function g_i describes how TFs regulate the transcription of gene i, and experimental evidence suggests that the response of mRNA to TF concentrations has a nonlinear Hill curve form [4]. Thus, the regulation function of transcription factor p_j on its target gene i can be described by

$$g_i^+ = v_i \frac{p_j^{h_{ij}}}{k_{ij}^{h_{ij}} + p_j^{h_{ij}}} \tag{3.6}$$

for the activation case and

$$g_i^- = v_i \frac{k_{ij}^{h_{ij}}}{k_{ij}^{h_{ij}} + p_j^{h_{ij}}} \tag{3.7}$$

for the inhibition case, where v_i is the maximum rate of transcription of gene i, k_{ij} is the concentration of protein p_j at which gene i reaches half of its maximum transcription rate and h_{ij} is a steepness parameter describing the shape of the nonlinear sigmoid responses.

Finally, nonlinear dynamics are also ubiquitous at the inter-cellular level, where linear models obviously cannot capture the saturation effects arising from limitations on the number of cells which can exist in a medium or organism. In fact, interactions between different types of cells often display highly nonlinear dynamics — consider, for example, a recently proposed (and validated) model of tumour-immune cell interactions, [5], which gives the relationships between tumour cells T, Natural Killer (NK) cells N and tumour-specific CD8+ T-cells L as,

$$\frac{dT}{dt} = aT(1 - bT) - cNT - D$$

$$\frac{dN}{dt} = \sigma - fN + \frac{gT^2}{h + T^2}N - pNT$$

$$\frac{dL}{dt} = -mL + \frac{jD^2}{k + D^2}L - qLT + rNT$$

$$D = d\frac{(L/T)^\gamma}{s + (L/T)^\gamma}T$$

Again, the highly nonlinear nature of the dynamics governing the interactions between the different cell types is strikingly apparent in the above equations.

For the reasons discussed above, the mathematical models which have been developed to describe the dynamics of intra- and inter-cellular networks are typically composed of sets of nonlinear differential equations, i.e. nonlinear dynamical systems. The study of such systems is by now a mature and extensive field of research in its own right (see, for example, [6]) and so we will not attempt to provide a complete treatment here. Instead, and in keeping with the aims of this book, we will focus on certain aspects of nonlinear systems and control theory which have particular applicability to the study of biological systems.

3.2 Equilibrium points

A biological system often operates in the neighbourhood of some nominal condition, i.e. the production and degradation rates of the biochemical compounds are regulated so that the amounts of each species remain approximately constant at some levels. When such an *equilibrium* is perturbed by an unpredicted event (e.g. by the presence of exogenous signalling molecules, like growth factors), a variety of different reactions may take place, which in general can lead the system either to operate at a different equilibrium point or to tackle the cause of the perturbation in order to restore the nominal operative condition.

A point x_e in the state-space of a generic nonlinear system* without exogenous inputs

$$\dot{x} = f(x) \qquad (3.8)$$

is said to be an *equilibrium point* if, whenever the state of the system starts at x_e, it will remain at x_e for all $t > 0$. The equilibrium points are the roots of the equation $f(x) = 0$. When the system has an exogenous input the generic model reads

$$\dot{x} = f(x, u) \qquad (3.9)$$

and the pair (x_e, u_e) is an equilibrium point for the system if

$$f(x_e, u_e) = 0 \, .$$

One of the main differences between linear and nonlinear systems is that the latter can exhibit zero, one or multiple isolated equilibria, which are in general different from the origin of the state-space. In the linear case, instead, the equation $Ax = 0$ admits only the trivial isolated solution $x = 0$, if $\det A \neq 0$, or a continuum of equilibrium points (e.g. a straight line in the state-space of a second order system), when A has one or more zero eigenvalues.

Example 3.1

Let us consider the basic reaction

$$R + S \underset{k_{off}}{\overset{k_{on}}{\rightleftharpoons}} C, \qquad (3.10)$$

which describes the reversible binding of a ligand S to a receptor molecule R with the formation of a complex C, with the binding and unbinding reaction rates given by the kinetic constants k_{on} and k_{off}, respectively. Applying the law of mass action, it is straightforward to write the ODE model of the above reaction as

$$\frac{dR}{dt} = \frac{dL}{dt} = k_{off}C - k_{on}R\,L \qquad (3.11a)$$

$$\frac{dC}{dt} = k_{on}R\,L - k_{off}C \qquad (3.11b)$$

---

*In the following we make certain mild assumptions about the mathematical properties of f (e.g. it is autonomous, piecewise continuous and locally Lipschitz, [6]) which are in practice true for the vast majority of models used in systems biology.

The equilibrium point of the reaction can be found by setting to zero all the derivatives of the state variables, i.e. by computing the solution of[†]

$$k_{\text{off}} C - k_{\text{on}} R\, L = 0$$

which yields

$$\frac{C_{\text{eq}}}{R_{\text{eq}}\, L_{\text{eq}}} = \frac{k_{\text{on}}}{k_{\text{off}}} = K_{\text{eq}}. \qquad (3.12)$$

The larger the *equilibrium constant* K_{eq}, the stronger the ligand–receptor binding; indeed, in the case of binding reactions, this is also called the *binding constant* and denoted by K_B (biochemists often refer also to the *dissociation constant*, defined as $K_D = k_{\text{off}}/k_{\text{on}} = 1/K_B$). The equilibrium constant is directly related to the *biochemical standard free energy change* $\Delta G'^{\circ}$, which gives the free energy change for the reacting system at standard conditions (temperature 298 K, partial pressure of each gas 1 atm, concentration of each solute 1 M, pH 7), by the expression [7]

$$\Delta G'^{\circ} = -RT \ln K'_{\text{eq}}, \qquad (3.13)$$

where R is the gas constant, 8.315 J/mol·K, and T is the absolute temperature. Thus, if the state of the system is at $(R_{\text{eq}},\ L_{\text{eq}},\ C_{\text{eq}})$, in the absence of exogenous perturbations the concentrations of the three species will remain at those values indefinitely. Note carefully, however, that this is a *dynamic equilibrium*: the system is not static; indeed, reactions are continuously taking place. However the binding and unbinding events per unit time are balanced such as to keep the overall concentrations unchanged. ⬜

It is fundamental to understand that the equilibrium points of a system depend not just on the structure of the equations, but also on the values of the parameters: in nonlinear systems, even small changes in the value of a single parameter can significantly alter the map of equilibrium points, as illustrated by the next example.

Example 3.2
Consider the nonlinear system

$$\frac{dx}{dt} = rx\left(1 - \frac{x}{q}\right) - \frac{x^2}{1 + x^2} = f(x) \qquad (3.14)$$

The equilibrium points are solutions of the equation $f(x) = 0$ which implies

$$rx\left(1 - \frac{x}{q}\right) - \frac{x^2}{1 + x^2} = 0$$

[†]This is a special case, because the resulting equations are all equivalent; in general, setting to zero the derivatives yields a number of algebraic equations equal to the number of state variables.

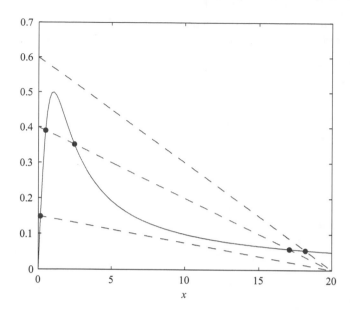

FIGURE 3.1: Intersections of $x/(1+x^2)$ (solid line) and $r(1-x/q)$ (dashed lines) for $q = 20$ and $r = 0.15, 0.4, 0.6$.

Obviously, $x = 0$ is an equilibrium point, but so are all the solutions to the equation

$$r\left(1 - \frac{x}{q}\right) = \frac{x}{1+x^2} \qquad (3.15)$$

These solutions can be easily visualised by plotting both sides of Eq. (3.15), as shown in Fig. 3.1: the intersections correspond to the equilibrium points of system (3.14). Note how both the location and the number of equilibrium points change for different values of the parameter r. ⬚

3.3 Linearisation around equilibrium points

The study of the behaviour of nonlinear systems is typically based on the computation of the equilibrium points and the subsequent analysis of the trajectories of the system in the neighbourhood of these points. Such an analysis can be conducted quite easily by computing linearised models that approximate the system behaviour around a given equilibrium point.

From a mathematical point of view, the linear approximation is based on the Taylor series expansion of $f(x)$ in the neighbourhood of x_e, that in the scalar case reads

$$f(x) = f(x_e) + \frac{df(x_e)}{dx}(x - x_e) + \frac{1}{2!}\frac{d^2 f(x_e)}{dx^2}(x - x_e)^2 + \frac{1}{3!}\frac{d^3 f(x_e)}{dx^3}(x - x_e)^3 + \dots$$

(3.16)

Hence, the linear (or first-order) approximation

$$f_a(x) = f(x_e) + \frac{df(x_e)}{dx}(x - x_e)$$

is actually close to $f(x)$ if at least one of these conditions holds:

a) The value of $(x - x_e)$ is small, i.e. the system trajectory is always close to x_e, and thus the terms $(x - x_e)^n$ with $n > 1$ are negligible;

b) The values of the derivatives $d^n f(x)/dx^n$ are negligible for $n > 1$, i.e. the function $f(x)$ is only 'mildly' nonlinear in the neighbourhood of x_e.

Similar arguments apply when the system's dimension is greater than one, in which case

$$f(x) : x \in \mathbb{R}^n \mapsto \begin{pmatrix} f_1(x) \\ \vdots \\ f_n(x) \end{pmatrix} \in \mathbb{R}^n$$

is a vector function. Therefore, in place of the derivative, we will use the *Jacobian*, that is the matrix of all first-order partial derivatives, defined as

$$J(x) = \frac{\partial f(x)}{\partial x} = \begin{pmatrix} \frac{\partial f_1}{\partial x_1} & \cdots & \frac{\partial f_1}{\partial x_n} \\ \vdots & \ddots & \vdots \\ \frac{\partial f_n}{\partial x_1} & \cdots & \frac{\partial f_n}{\partial x_n} \end{pmatrix}.$$

Thus, the behaviour of the nonlinear system (3.8) can be approximated in the neighbourhood of an equilibrium point x_e by the linear system

$$\delta\dot{x} = J(x_e)\delta x,$$

(3.17)

where $\delta x = x - x_e$.

Example 3.3

The second-order system

$$\dot{x}_1 = f_1(x_1, x_2) = \nu_0 + \nu_1\beta - \nu_2 + \nu_3 + k_f x_2 - k x_1 \qquad (3.18a)$$
$$\dot{x}_2 = f_2(x_1, x_2) = \nu_2 - \nu_3 - k_f x_2 \qquad (3.18b)$$

devised in [8] defines a minimal model for intracellular Ca^{2+} oscillations induced by the rise of inositol 1,4,5-trisphosphate (InsP3), which is triggered by external signals that bind to the cell membrane receptor phospholipase C (PLC). The increase of $InsP_3$ concentration triggers the release of Ca^{2+} from intracellular stores. Subsequently, Ca^{2+} oscillations arise from the cyclic exchange of this ion between the cytosol and a pool insensitive to $InsP_3$.

The two state variables in this model are the concentrations of free Ca^{2+} in the cytosol (x_1) and in the $InsP_3$-insensitive pool (x_2). The terms ν_0 and kx_1 represent the influx and efflux of Ca^{2+} into and out of the cell, respectively, in the absence of external stimuli. The rates of Ca^{2+} transport from the cytosol into the $InsP_3$-insensitive store and vice versa are

$$\nu_2 = V_{M2} \frac{x_1^n}{K_2^n + x_1^n} \tag{3.19}$$

$$\nu_3 = V_{M3} \frac{x_2^m}{K_R^m + x_2^m} \cdot \frac{x_1^p}{K_A^p + x_1^p} \tag{3.20}$$

Furthermore, there is a nonactivated, leaky transport of x_2 into x_1, given by $k_f x_2$. In the model, the level of $InsP_3$ is assumed to affect the influx of Ca^{2+} by raising the value of β. Parameter values for the model are shown in Table 3.1. In order to derive a linearised version of the above model we calculate the partial derivatives

$$\frac{\partial f_1}{\partial x_1} = -\frac{130\,x_1}{\left(x_1^2 + 1\right)^2} + \frac{1.312\mathrm{e}3\,x_1^3\,x_2^2}{\left(x_1^4 + 0.6561\right)^2\left(x_2^2 + 4\right)} - 10 \tag{3.21a}$$

$$\frac{\partial f_1}{\partial x_2} = \frac{4\mathrm{e}3\,x_1^4\,x_2}{\left(x_2^2 + 4\right)^2\left(x_1^4 + 0.6561\right)} + 1 \tag{3.21b}$$

$$\frac{\partial f_2}{\partial x_1} = \frac{130\,x_1}{\left(x_1^2 + 1\right)^2} - \frac{1.312\mathrm{e}3\,x_1^3\,x_2^2}{\left(x_2^2 + 4\right)\left(x_1^4 + 0.6561\right)^2} \tag{3.21c}$$

$$\frac{\partial f_2}{\partial x_2} = -\frac{4\mathrm{e}3\,x_1^4\,x_2}{\left(x_2^2 + 4\right)^2\left(x_1^4 + 0.6561\right)} - 1 \tag{3.21d}$$

$$\tag{3.21e}$$

Note that the Jacobian is independent of β, since the term $\nu_1\beta$ in Eq. (3.18a) does not depend on x. The value of β, instead, determines the location of the equilibrium point. Assuming $\beta = 0.23$, the linearised system is derived by computing the Jacobian at the equilibrium point, i.e. by setting $x = x_e = (0.2679, 2.2108)$ in Eq. (3.21), which yields

$$J(x_e) = \begin{pmatrix} -8.5872 & 1.8721 \\ -1.4128 & -1.8721 \end{pmatrix}. \tag{3.22}$$

The free evolution of the nonlinear system (3.18) and that of its linearisation, with initial condition $(0.18, 2)$, are compared in Fig. 3.2: the trajectories are very similar since they are close to the equilibrium point. Fig. 3.3, on the

TABLE 3.1
Kinetic parameters for the Goldbeter model (3.18)

Parameter	Value	Unit	Parameter	Value	Unit
ν_0	1	$\mu M \cdot s^{-1}$	K_2	1	μM
k	10	s^{-1}	K_R	2	μM
k_f	1	s^{-1}	K_A	0.9	μM
ν_1	7.3	$\mu M \cdot s^{-1}$	m	2	
V_{M_2}	65	$\mu M \cdot s^{-1}$	n	2	
V_{M_3}	500	$\mu M \cdot s^{-1}$	p	4	

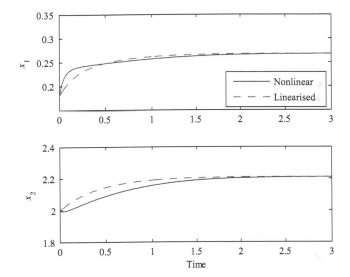

FIGURE 3.2: Free evolution state response of the nonlinear system (3.18) and of its linearisation, with initial condition $(0.18, 2)$ close to the equilibrium point.

other hand, shows what happens when the trajectories start from a point that is further away from the equilibrium, that is $(0.42, 1.85)$. □

Linearisation can be applied also in the presence of exogenous inputs, e.g. for system (3.9) the linearised system is

$$\delta \dot{x} = \begin{pmatrix} \frac{\partial f_1}{\partial x_1} & \cdots & \frac{\partial f_1}{\partial x_n} \\ \vdots & \ddots & \vdots \\ \frac{\partial f_n}{\partial x_1} & \cdots & \frac{\partial f_n}{\partial x_n} \end{pmatrix} \delta x + \begin{pmatrix} \frac{\partial f_1}{\partial u_1} & \cdots & \frac{\partial f_1}{\partial u_n} \\ \vdots & \ddots & \vdots \\ \frac{\partial f_n}{\partial u_1} & \cdots & \frac{\partial f_n}{\partial u_n} \end{pmatrix} \delta u \qquad (3.23)$$

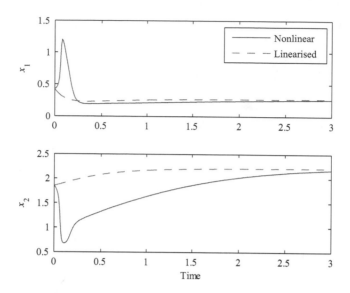

FIGURE 3.3: Free evolution state response of the nonlinear system (3.18) and of its linearisation, with initial condition $(0.42, 1.85)$ far from the equilibrium point.

where $\delta u = u - u_e$.

Example 3.4

In the model introduced in the previous example, if β is considered a variable exogenous input, we end up with the following linearised system around an equilibrium (x_e, β_e)

$$\delta \dot{x} = \begin{pmatrix} -8.5872 & 1.8721 \\ -1.4128 & -1.8721 \end{pmatrix} \delta x + \begin{pmatrix} 7.3 \\ 0 \end{pmatrix} \delta \beta, \qquad (3.24)$$

where $\delta \beta = \beta - \beta_e$. From Figs. 3.4 and 3.5 it is possible to see that for small step inputs the responses of the nonlinear system and of its linearisation are very close, whereas they are completely different when the amplitude of the step input increases.   □

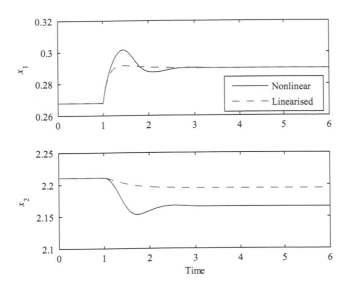

FIGURE 3.4: State response of the nonlinear system (3.18) and of its linearisation to a small step change of β (from 0.23 to 0.26), applied at $t = 1$.

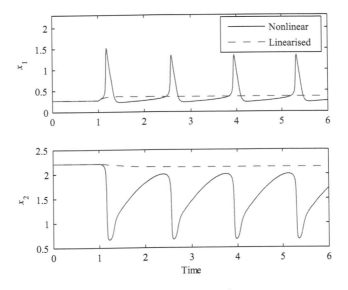

FIGURE 3.5: State response of the nonlinear system (3.18) and of its linearisation to a large step change of β (from 0.23 to 0.35), applied at $t = 1$.

3.4   Stability and regions of attractions

Since the behaviour of a nonlinear system in a small neighbourhood of an
equilibrium point is usually well approximated by its linearisation, the question that naturally arises is how to guarantee that the state trajectories do not
deviate far from an equilibrium point after the system is subject to a small
perturbation: this leads us to the concept of *stability*. Roughly speaking, an
equilibrium point x_e of a system (3.8) is stable if all the system's trajectories
starting from a small neighbourhood of x_e stay close to x_e for all time.

Considering the above, it comes as no surprise that stability is among the
most important and thus the most investigated properties in control and dynamical systems theory. It is also worth noting that, besides state-space equilibrium points, other kinds of stability can be considered, e.g. input-output
stability and stability of periodic orbits (i.e. limit cycles).

3.4.1   Lyapunov stability

Many of the fundamental results concerning the stability of dynamical systems
are due to the work of the Russian mathematician Lyapunov, who devised the
homonymous stability theory. The formal definition of Lyapunov stability is
given below in the case when the equilibrium coincides with the origin of the
state-space[‡].

Stability of an equilibrium point: *The origin of the state-space*
$x = 0$ *is a stable equilibrium point at $t = 0$ for system* (3.8) *if for each*
$\epsilon > 0$ *and any $t_0 \geq 0$ there is a $\delta > 0$ such that*

$$\|x(t_0)\| < \delta \Rightarrow \|x(t)\| < \epsilon, \quad t \geq t_0 .$$

This means that, if we choose an initial condition which is close enough
to the equilibrium point, the trajectory of the system is guaranteed not to
drift away by more than a specified distance from the equilibrium point (see
Fig. 3.6). Moreover, if the trajectory tends asymptotically to the stable equilibrium point, that is

$$\|x(t_0)\| < \delta \Rightarrow \lim_{t \to \infty} x(t) = 0 ,$$

the equilibrium point is said to be *asymptotically stable*. In this case, all trajectories tend to the equilibrium point provided that the initial condition is
sufficiently close: the set of all such initial conditions is denoted the *Region* (or
Domain) *of Attraction* of the equilibrium point. The analytical computation

---

[‡]It can be easily shown that the general case of nonzero equilibrium can be recast in the
same form by applying a change of variables.

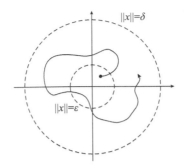

FIGURE 3.6: Stability of an equilibrium point.

of this region is usually not an easy task, but it can often be either computed numerically or approximated by using the level curves of some suitable function.

Lyapunov's theory also provides us with some useful tools with which to analyse the stability of an equilibrium point, for a given system of nonlinear differential equations.

Lyapunov's Direct Method: *Let $x = 0$ be an equilibrium point for system (3.8) and $D \subset \mathbb{R}^n$ be a domain containing $x = 0$. Let $V : D \to \mathbb{R}$ be a continuously differentiable function such that*

$$V(0) = 0 \tag{3.25a}$$

$$V(x) > 0, \quad \forall x \in D - \{0\} \tag{3.25b}$$

$$\dot{V}(x) \leq 0, \quad \forall x \in D. \tag{3.25c}$$

Then, $x = 0$ is stable. Moreover, if

$$\dot{V}(x) < 0, \quad \forall x \in D - \{0\},$$

then $x = 0$ is asymptotically stable.

A continuously differentiable function $V(x)$ satisfying Eq. (3.25) is called a *Lyapunov function*. A significant advantage of the above result is that it allows us to determine the stability of an equilibrium point without needing to actually compute the trajectories of the system. On the other hand, the theorem provides only a sufficient condition for stability and, therefore, if one fails in finding a suitable Lyapunov function, one cannot conclude that the equilibrium point is unstable.

One of the main difficulties in applying Lyapunov's direct method is that there is no systematic way to construct a suitable solution $V(x)$. The most common approach for generating a candidate Lyapunov function is to construct a quadratic function having the form

$$V(x) = x^T P x, \qquad (3.26)$$

where P is a square symmetric matrix. For this form, it is quite straightforward to determine positive or negative definiteness, as this corresponds to all of the eigenvalues of P being either positive or negative, in each case.

Example 3.5

In Example 3.3 we have determined that $(0.2679, 2.2108)$ is an equilibrium point for the system when $\beta = 0.23$. Now we want to use Lyapunov's direct method to analyse the stability of this equilibrium point. Taking a quadratic Lyapunov function centered on the equilibrium point, that is

$$V(x) = (x - x_e)^T P (x - x_e),$$

the conditions of Eq. (3.25) are verified if

$$V(x) > 0, \quad \forall x \in \mathcal{D} - \{x_e\}$$
$$\dot{V}(x) = f(x)^T P(x - x_e) + (x - x_e)^T P f(x) \leq 0, \quad \forall x \in \mathcal{D},$$

for some domain $\mathcal{D} \subset \mathbb{R} : x_e \in \mathcal{D}$. The matrix

$$P = \begin{pmatrix} 0.5684 & -0.0634 \\ -0.0634 & 0.0746 \end{pmatrix}$$

is a solution to this problem (in Section 3.5.3 we will show how this matrix was found); indeed, the surfaces shown in Figs. 3.7–3.8 show that there exists a neighbourhood of x_e where the Lyapunov function is positive, and its derivative is strictly negative. Therefore, we can conclude that the equilibrium point x_e is asymptotically stable. ☐

Another, in many cases simpler, method to check the stability of equilibrium points is the so-called Lyapunov's indirect method, which is stated below.

Lyapunov's Indirect Method: *Let $x = 0$ be an equilibrium point for the nonlinear system (3.8), where $f : \mathcal{D} \to \mathbb{R}^n$ is continuously differentiable and $\mathcal{D}$ is a neighbourhood of the origin and let*

$$A = \left. \frac{\partial f}{\partial x}(x) \right|_{x=0}.$$

Then

a) *The origin is asymptotically stable if $\Re(\lambda_i) < 0$ for all the eigenvalues λ_i, $i = 1, \dots, n$ of A.*

b) *The origin is unstable if $\Re(\lambda_i) > 0$ for one or more eigenvalues of A.*

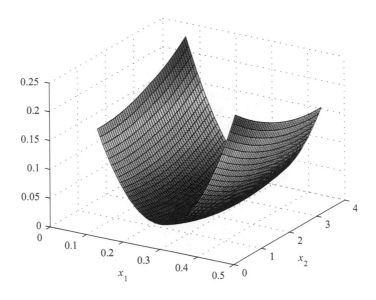

FIGURE 3.7: Surface described by the quadratic Lyapunov function in Example 3.5.

Example 3.6

Consider again the stability of the equilibrium point x_e of the system (3.18). In Example 3.4 we computed the linearisation of this system around x_e and the Jacobian, $J(x_e)$, is given in Eq. (3.22). To apply Lyapunov's indirect method, it is sufficient to compute the eigenvalues of $J(x_e)$, which are -8.167 and -2.293: since they are both strictly negative, we can conclude that the equilibrium point is asymptotically stable. ▯

Note that Lyapunov's indirect method, like the direct one, provides only sufficient conditions for stability. It also does not allow us to determine the stability of an equilibrium when there are one or more eigenvalues with zero real part (i.e. $\Re(\lambda_i) = 0$ for some i).

3.4.2 Region of attraction

Regions of attractions are of paramount importance in biological systems; indeed, when a system that is operating in the neighbourhood of an equilibrium point leaves the boundaries of its region of attraction, it is usually abruptly led to a new operating condition (corresponding to another equilibrium point or to oscillations). This phenomenon is at the basis of many on-off regulatory mechanisms of biological functions and periodic behaviours, at the molecular,

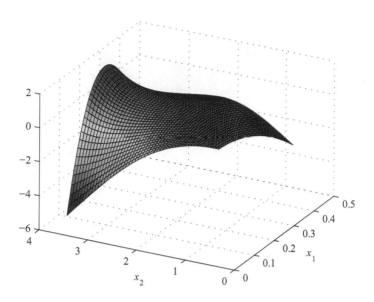

FIGURE 3.8: Surface described by the derivative of the Lyapunov function in Example 3.5.

cellular or population level. First of all let us give a more precise definition of region of attraction.

> **Region of Attraction:** *Let the origin $x = 0$ be an asymptotically stable equilibrium point for the nonlinear system (3.8) and let $\Phi(t, x_0)$ be the solution of Eq. (3.8) that starts at initial state x_0 at time $t = 0$. The region of attraction of the origin, denoted by $\mathcal{R}_A$, is defined by*

$$\mathcal{R}_A = \{x_0 : \Phi(t, x_0) \to 0 \text{ as } t \to \infty\}.$$

A notable property of the region of attraction is that the boundary of $\mathcal{R}_A$ is always formed by trajectories of the system. This makes it quite easy to numerically determine, for second-order systems, the region of attraction on the *phase plane*. The phase plane is a diagram giving the trajectories of a second-order system, that is the curve described by the point $(x_1(t), x_2(t))$ as the time t varies.

Example 3.7

Consider the hypothetical genetic regulatory system depicted in Fig. 3.9, composed of two genes, G_1 and G_2, whose expression is regulated by the corresponding proteins, P_1 and P_2. Each gene promotes its own transcription and inhibits that of the other gene.

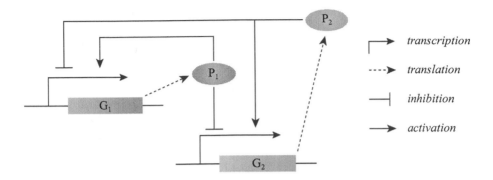

FIGURE 3.9: A network of two genes with self and mutual trascriptional regulation.

Denote the concentrations of the mRNA molecules transcribed from G_1 and G_2 as x and y respectively. In order to derive a second-order system, we will neglect the dynamics of mRNA translation into the corresponding proteins. Then, the system dynamics can be modelled as

$$\dot{x} = \mu_1 \left(\frac{\alpha_0 + \alpha_1 \left(\frac{x}{K_1}\right)^{h_1}}{1 + \left(\frac{x}{K_1}\right)^{h_1} + \left(\frac{y}{K_2}\right)^{h_2}} \right) - \lambda_{d1} x \qquad (3.27a)$$

$$\dot{y} = \mu_2 \left(\frac{\beta_0 + \beta_1 \left(\frac{y}{K_3}\right)^{h_3}}{1 + \left(\frac{x}{K_4}\right)^{h_4} + \left(\frac{y}{K_3}\right)^{h_3}} \right) - \lambda_{d2} y. \qquad (3.27b)$$

The regulatory terms in parentheses express the combinatorial effects of the two species on the mRNA transcription. Note that, when the concentrations of the two proteins are zero, the genes are transcribed at the basal rates ($\mu_1 \alpha_0$ and $\mu_2 \beta_0$, respectively). Choosing the parameter values[§] given in Table 3.2, the system exhibits five equilibrium points, whose values are shown in Table 3.3. Using Lyapunov's indirect method, we can readily establish that there are three asymptotically stable equilibrium points: one corresponds to low expression levels for both genes; the other two occur when one of the two genes is highly expressed and the other is almost knocked out. Therefore,

---

[§]The parameters given in Table 3.2 have been arbitrarily chosen, and should not be assumed to be representative of experimental concentrations and kinetic constants.

TABLE 3.2
Kinetic parameters for the two-genes regulatory network in Example 3.7

Parameter	Value	Unit	Parameter	Value	Unit
μ_1	0.1	$\mu M \cdot s^{-1}$	α_0	0.1	
μ_2	0.1	$\mu M \cdot s^{-1}$	α_1	0.9	
λ_{d1}	0.1	s^{-1}	β_0	0.1	
λ_{d2}	0.1	s^{-1}	β_1	0.9	
K_1	0.5	μM	h_1	3	
K_2	1	μM	h_2	3	
K_3	0.5	μM	h_3	3	
K_4	1	μM	h_4	3	

TABLE 3.3
Equilibrium points and eigenvalues of the linearisation of system (3.27)

Equilibrium point	Eigenvalues
(0.0814, 0.6483)	-0.089861, -0.020127
(0.0935, 0.5010)	-0.08509, 0.019804
(0.1078, 0.1078)	-0.077778, -0.078521
(0.5010, 0.0935)	0.019804, -0.08509
(0.6483, 0.0814)	-0.020127, -0.089861

from an engineering perspective, the system implements a tri-stable switch and the state can be controlled by some external effectors acting on one or both the genes to move it from one equilibrium point to another. This is effectively shown by the phase plane diagram in Fig. 3.10, where the (unbounded) regions of attraction, $\mathcal{D}_1$, $\mathcal{D}_2$, $\mathcal{D}_2$, of the three stable equilibrium points are separated by the dashed trajectories.   ▯

A practical means to estimate the region of attraction of an equilibrium point is provided by Lyapunov's results. First of all, it is important to remark that the domain $\mathcal{D}$ given above in the statement of Lyapunov's direct method cannot be considered an estimate of the region of attraction. Indeed, the region of attraction has to be an invariant set, that is every trajectory starting inside it should never escape from it. Since the Lyapunov function is positive with negative derivative over $\mathcal{D}$, the trajectory $x(t)$ is forced to move from greater to smaller level curves, defined by $V(x) = x^T P x = c$, where c is a positive scalar. This, however, does not prohibit the trajectory from trespassing the boundary of $\mathcal{D}$, thus possibly leading the state to a different equilibrium

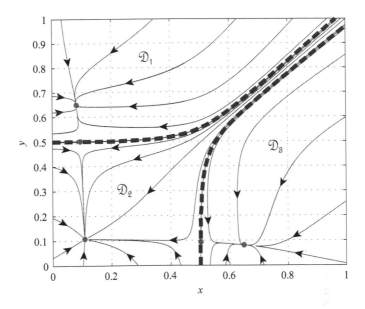

FIGURE 3.10: Phase plane of system (3.27).

point or to diverge. Therefore, the best estimate that one can provide using Lyapunov's results is represented by the largest invariant set contained in $\mathcal{D}$. A simple way to compute such an estimate is to find the maximum value of c such that the corresponding Lyapunov function level curve, $V(x) = c$, is completely included in $\mathcal{D}$, [6].

3.5   Optimisation methods for nonlinear systems

Modern control engineering makes widespread use of mathematical optimisation to design and analyse feedback control systems. As the scale and complexity of industrial control systems has increased over recent years, control engineers have been forced to abandon many traditional linear design techniques in favour of nonlinear methods which often rely on numerical optimisation. Increasingly, the emphasis is on finding ways to formulate optimisation problems which accurately reflect a set of design or analysis criteria and are tractable in terms of the computational requirements of the corresponding optimisation algorithms. In systems biology, the ever increasing scale of the nonlinear computational models being developed has also highlighted the important role of optimisation in developing, validating and interrogating these models, [9]. Some of the tasks for which advanced optimisation methods are now widely used in systems biology research include

- Model parameter estimation against experimental data for "bottom-up" modelling, [10, 11]

- Network inference in "top-down" modelling, [12, 13]

- Analysing model stability and robustness, [14, 15]

- Exploring the potential for yield optimisation in biotechnology and metabolic engineering, [16, 17]

- Directly optimising experimental searches for optimal drug combinations, [18]

- Computational design of genetic circuits, [19]

- Optimal control for modification of self-organised dynamics, [20]

- Optimal experimental design, [21]

In many of the above examples, the analysis of the particular system property of interest can be formulated as an optimisation problem of the form

$$\max_x f(x) \text{ subject to } y \leq x \leq z$$

or

$$\min_x f(x) \text{ subject to } y \leq x \leq z$$

where x is a vector of model parameters with upper and lower bounds z and y, respectively, and $f(x)$ is some nonlinear *objective function* or *cost function*. For example, in a model validation problem, the objective function could be formulated as the difference between the simulated outputs of the model and one or more sets of corresponding experimental data, and the optimisation algorithm would compute the values of the model parameters within their biologically plausible ranges (defined by y and z) which minimise this function. On the other hand, in a yield optimisation problem $f(x)$ could represent the total quantity of some product produced in a given time period. The optimisation algorithm would then search for parameter values and/or model structures which maximised this quantity in the model simulation, in order, for example, to provide guidance on the choice of mutant strains for improving yields.

Many different classes of optimisation algorithms are available in the literature. In this section, we give a brief overview of some of the most widely used optimisation methods which may be used to solve problems of the type considered in systems biology research. The application of several of these methods to particular problems in systems biology will be illustrated in the two Case Studies at the end of this chapter and in later chapters.

3.5.1 Local optimisation methods

Local algorithms typically make use of gradient information (calculated either analytically of numerically) of the cost function to find the search direction while determining the optimum. Global algorithms, in contrast, typically use randomisation and/or heuristic search techniques which require only the calculation of the objective function value. The search space, or design space, for the set of optimisation parameters being used may be convex or non-convex. Fig. 3.11 shows a two-dimensional convex search space in the parameters x and y, with a corresponding cost function z. Clearly, in this case there exists only one maximum value of the cost function over the entire search space, and thus any local optimisation algorithm which uses gradient information will eventually converge to the desired global solution. On the other hand, for

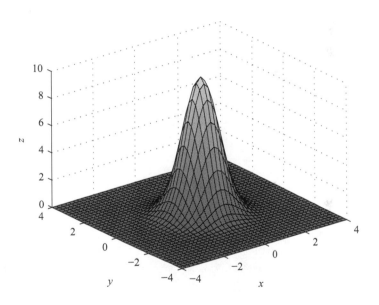

FIGURE 3.11: Example of a convex search space.

problems with non-convex search spaces, such as the one shown in Fig. 3.12, for example, gradient-based local optimisation algorithms may only provide a local, rather than a global solution, depending on where in the search space the optimisation starts. The performance of a given optimisation algorithm is generally problem dependent, and there is no unique optimisation algorithm for general classes of problems which will guarantee computation of the true global solution with reasonable computational complexity. One of the most

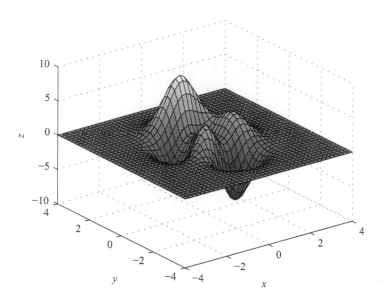

FIGURE 3.12: Example of a non-convex search space.

widely used local optimisation algorithms is the Sequential Quadratic Progamming (SQP) method, which is often a very effective approach for medium-size non-linearly constrained optimisation problems. It can be seen as a generalisation of Newton's method for unconstrained optimisation in that it finds a step away from the current point by minimising from a sequence of quadratic programming subproblems. SQP methods are efficient general purpose algorithms for solving smooth and well-scaled non-linear optimisation problems when the functions and gradients can be evaluated with high precision. In many situations, the local gradients will not be available analytically and in such situations numerical approximations of gradients have to be computed — this can cause slower and less reliable performance, especially when the function evaluations are noisy. In SQP, a quadratic approximation of the Lagrange function and an approximation of the Hessian matrix are defined by a quasi-Newton matrix. The SQP algorithm replaces the objective function with a quadratic approximation and replaces the constraints with linear approximations. The quasi-Newton matrix is updated in every iteration using the standard Broyden-Fletcher-Goldfarb-Shanno (BFGS) formula. An efficient MATLAB® coding of the SQP algorithm is available as the function *"fmincon"* provided in [22]; the associated documentation also provides more details on the SQP method.

3.5.2 Global optimisation methods

Most optimisation problems encountered in systems biology involve non-convex search spaces and thus require the use of global optimisation methods to ensure the computation of globally optimal solutions. Global optimisation algorithms may be broadly separated into two classes: evolutionary algorithms and deterministic search methods. The most well-known type of evolutionary algorithms are Genetic Algorithms (GA's), which are general purpose stochastic search and optimisation algorithms, based on genetic and evolutionary principles [23]. This approach assumes that the evolutionary processes observed in nature can be simulated on a computer to generate a *population*, or a set, of fittest candidates. A fitness function (corresponding to the objective function of interest) is defined to assign a performance index to each candidate. In genetic search techniques, each member of the population of candidates is encoded as an artificial chromosome, and the population undergoes a repetitive evolutionary process of reproduction through *selection* for mating according to a fitness function and recombination via *crossover* with *mutation*. A complete repetitive sequence of these genetic operations is called a *generation*. GA's have become a popular, robust search and optimisation technique for problems with large as well as small parameter search spaces. Due to their stochastic nature, global optimisation schemes such as GA's can be expected to have a much better chance of converging to a global optimum than local optimisation algorithms. The price to be paid for this improved performance is a dramatic increase in computation time when compared with local methods. Fleming and Purshouse, in [24], provide a comprehensive review of various applications of GA's in the field of control engineering.

Differential evolution (DE) is a relatively new global optimisation method, introduced by Storn and Price in [25]. It belongs to the same class of evolutionary global optimisation techniques as GA's, but unlike GA's it does not require either a selection operator or a particular encoding scheme. Despite its apparent simplicity, the quality of the solutions computed using DE is claimed to be generally better than those achieved using other evolutionary algorithms, both in terms of accuracy and computational overhead [25, 26, 27]. This method also starts the optimisation from randomly generated multiple candidate solutions. In DE, however, a new search point in each iteration is generated by adding the weighted vector difference between two randomly selected candidate points in the population, with yet another third randomly chosen point. The vector difference determines the search direction and a weighting factor decides the step size in that particular search direction. The DE methodology consists of the following four main steps 1) Random initialisation, 2) Mutation, 3) Crossover, 4) Evaluation and Selection. There are different schemes of DE available based on the various operators that are employed; one of the most popular is referred to as "DE/rand/1/bin" [25].

A significant drawback of the evolutionary methods described above is that no formal proofs of convergence are available, and hence multiple trials may be required to provide confidence that the global solution has been found. An alternative approach is to use deterministic methods such as the DIRECT (DIviding RECTangles) algorithm, [28], a modified version of a class of Lipschitzian optimisation schemes, which, when run for a long enough time, has been proved to converge to the global solution [29]. The DIRECT algorithm has previously been successfully applied to several different classes of optimisation problems. In [30], DIRECT optimisation is applied to a realistic slider air-bearing surface (ABS) design, an important engineering optimisation problem in magnetic hard disk drives, in which the cost function evaluation also requires substantial computational time. Fast convergence of the algorithm and a favourable comparison with adaptive simulated annealing were demonstrated in this study. In [31] the minimisation of the cost of fuel and/or electric power for the compressor stations in a gas pipeline network is attempted using the DIRECT algorithm and a hybrid version of the DIRECT algorithm with implicit filtering. Again, the application is a complex and realistic one, and the reported results are very promising. In [10] the DIRECT optimisation method is used to improve the set of estimated parameters in a model of mitotic control in frog egg extracts. DIRECT optimisation is used to search for the globally optimal kinetic rate constants for a proposed mathematical model of the control system that best fits the experimental data set, and the improvement obtained over the locally optimised parameter set was clearly demonstrated.

Whatever the particular problem under consideration, the highly complex and nonlinear nature of biological systems' dynamics means that the search space is often non-convex and of high dimension, with computationally expensive objective function evaluations. In order to gain confidence (with reasonable computational overheads) that the global optimum for the problem has been found, it is often useful to employ two algorithms (say DE and DIRECT) which are based on completely different principles and strategies, and check whether the results are consistent. In addition, combining the best features of global and local methods in *hybrid* algorithms can sometimes produce significant computational savings as well as improved performance, [32, 33]. In such hybrid schemes there is the possibility of incorporating domain knowledge, which gives them an advantage over a pure blind search based on evolutionary principles such as GA's. Most of these hybrid schemes apply a technique of switching from the global scheme to the local scheme after the first optimisation algorithm finishes its search or optimisation. In [33], some guidelines are provided on designing more sophisticated hybrid GA's based on probabilistic switching strategies, along with experimental results and supporting mathematical analysis. Efficient computer code implementations and further details of all the global optimisation algorithms discussed above are available in the MATLAB® Genetic Algorithm and Direct Search Toolbox, [34].

3.5.3 Linear matrix inequalities

Linear Matrix Inequalities (LMI) play a crucial role in systems and control theory; indeed, they appear in the solutions to several important problems, e.g. construction of quadratic Lyapunov functions for stability and performance analysis, optimal control and interpolation problems. Their widespread application in the field arises from the development, in the late 1980's, of so called interior-point algorithms, which have proven to be extremely efficient methods for solving LMIs, enabling high-order problems to be tackled with standard computational requirements and in reasonable time.

An LMI is a particular type of *convex optimisation* problem, having the form

$$F(x) = F_0 + \sum_{i=1}^{n} x_i F_i > 0, \tag{3.28}$$

where the symmetric matrices F_i are assigned and x_i, $i = 1, \ldots, n$, are the optimisation variables. A noteworthy property of LMIs is that a set of multiple LMIs

$$F^{(1)} > 0, \ldots, F^{(p)} > 0$$

can be recast as the single LMI

$$\text{diag}(F^{(1)}, \ldots, F^{(p)}) > 0.$$

Moreover, some convex nonlinear inequalities can be converted to LMI form using the properties of Schur complements: the LMI

$$\begin{pmatrix} Q(x) & S(x) \\ S(x)^T & R(x) \end{pmatrix} > 0, \tag{3.29}$$

where $Q(x) = Q(x)^T$, $R(x) = R(x)^T$ and $S(x)$ depend affinely on x, is equivalent to

$$R(x) > 0, \quad Q(x) - S(x)R(x)^{-1}S(x)^T > 0.$$

In most cases, LMIs are encountered in a different form than Eq. (3.28), where the optimisation variables are arranged in a matrix format. For example, as we now show, the Lyapunov stability conditions of Eq. (3.31) are LMIs, where the optimisation variables are the entries of the symmetric matrix P.

Example 3.8

Consider again the system in Example 3.5, where the stability of the equilibrium point x_e was demonstrated by construction of the Lyapunov function

$$V(x) = (x - x_e)^T P(x - x_e).$$

This function satisfies the conditions for Lyapunov stability given in Eq. (3.25) if

$$V(x) > 0, \quad \forall x \in \mathcal{D} - \{x_e\}$$

$$\dot{V}(x) = f(x)^T P(x - x_e) + (x - x_e)^T P f(x) \leq 0, \quad \forall x \in \mathcal{D},$$

for some domain $\mathcal{D} \subset \mathbb{R} : x_e \in \mathcal{D}$. To find a matrix P which satisfies these conditions, we first apply the asymptotic stability conditions to the linearised system

$$\delta \dot{x} = A\,\delta x, \quad \text{where} \quad \delta x = x - x_e,$$

to get

$$V(\delta x) = \delta x^T P \delta x > 0 \tag{3.30a}$$

$$\dot{V}(\delta x) = \delta x^T P \delta \dot{x} + \delta \dot{x} P \delta x$$

$$= \delta x^T A^T P \delta x + \delta x^T P A \delta x < 0 \tag{3.30b}$$

Condition (3.30a) can be easily imposed by requiring the matrix P to be positive definite, that is

$$P > 0, \tag{3.31a}$$

whereas condition (3.30b) is satisfied by any matrix P that satisfies the linear matrix inequality

$$A^T P + P A < 0. \tag{3.31b}$$

The solution to the above set of LMIs can be calculated using standard optimisation packages (e.g. [35] or [36]) and gives

$$P = \begin{pmatrix} 0.5684 & -0.0634 \\ -0.0634 & 0.0746 \end{pmatrix}$$

This was the approach that was used to compute $V(x)$ in Example 3.5.   □

3.6 Case Study III: Stability analysis of tumour dormancy equilibrium

Biology background: Tumour progression is the process by which tumours grow and eventually invade surrounding tissues and/or spread (metastasise) to areas outside the local tissue. These metastatic tumours are the most dangerous and account for a large percentage of cancer deaths. The dynamics of tumour progression may be thought of in terms of a complex predator-prey system involving the immune system and cancer cells.

The immune system can recognise mutant or otherwise abnormal cells as foreign, but some cancer cells are able to mutate sufficiently that they are able to escape the surveillance mechanisms of the immune system. Certain cancers are able to produce chemical signals that inhibit the actions of immune cells, and some tumours grow in locations such as the eyes or brain, which are not regularly patrolled by immune cells. The population of "predators" thus consists of the immune response cells (white blood cells), such as T-lymphocytes, macrophages and natural killer cells: these cells engulf and neutralise malignant cells in a variety of ways.

Natural killer cells are cytotoxic — small granules in their cytoplasm contain special proteins such as perforin and proteases known as granzymes. When these are released in close proximity to a cell which has been earmarked for killing, perforin forms pores in the cell membrane of the target cell through which the granzymes and associated molecules can enter, inducing apoptosis.

Macrophages are another type of white blood cell that differentiate from blood monocytes which migrate into the tissues of the body. As well as helping to destroy bacteria, protozoa and tumour cells, macrophages also release substances that stimulate other cells of the immune system. T-lymphocytes originate in the bone marrow and reside in lymphatic tissue such as the lymph nodes and the spleen. T-lymphocytes are divided into two categories: regulatory and cytotoxic. In the regulatory form, *helper* T-lymphocytes organise the attack against the tumour cells (the prey), but they do not actively participate in the elimination of the malignant cells. They are able to stimulate the growth of the population of several types of predator cells (e.g. macrophages and cytotoxic T-lymphocytes). Moreover, predator cells are also present in the body in two forms: hunting and resting. For example, the cytotoxic T-lymphocytes in the resting form can become active predators (cytotoxic cells) when a helper T-cell sends an appropriate activation signal. Recent research has indicated that the immune system can also arrest development (induce dormancy) in early stage tumours without having to actually destroy the malignant cells.

3.6.1   Introduction

Intensive efforts have been made in recent systems biology research to develop reliable dynamical models of tumour development — see, for example, [37] and [38], which provide a comprehensive overview of different approaches to modelling of the tumour–immune system interaction dynamics. Recent work has indicated that functional models of competing populations (i.e. Lotka-Volterra-like models), in which tumour growth dynamics are explained in terms of competition between malignant and normal cells, provide many insights into the role of cell-cell interactions in growth regulation of tumours. In spite of their simple formulation, such models can capture many key features of cancer development, such as: *a)* unbounded growth, which leads to an uncontrolled tumour; *b)* steady-state conditions, in which the populations of normal and malignant cells coexist and their sizes do not vary (tumour dormancy); *c)* cyclic profiles of the size of the tumour cell population (tumour recurrence); and *d)* a steady-state of tumour eradication due to the action of the immune response (tumour remission). The cases *b)* and *d)* represent desirable clinical conditions since in these equilibria the population size of tumour cells can be constrained to low or null values.

In this Case Study, we consider the dynamical model of tumour growth devised in [39], which has three equilibrium points, two unstable, E_1, E_2, and one asymptotically stable, E_3. In the clinical context, E_1 and E_2 correspond to unbounded tumour growth, while E_3 corresponds to a density of malignant cells that can be considered safe for the patient and remains in a steady state (tumour dormancy) under the control of the immune system. As we will show, a nonlinear analysis of the tumour dynamics can determine an estimate for the region of attraction of the desirable equilibrium point, thus allowing us to map the range of clinical conditions under which the tumour progression can be kept under control through immune therapy.

The model considered in [39] belongs to a special class of nonlinear systems, namely quadratic systems, in which the nonlinearity arises from multiplicative terms between two state variables. Such systems arise in a vast array of applications in engineering (electrical power systems, chemical reactors, robots) as well as in ecological and biological systems, where the quadratic terms naturally arise when considering, for example, biochemical phenomena (e.g. from the law of mass action) or prey-predator-like interactions between multi-species populations.

The exact determination of the whole region of attraction of the zero equilibrium point of a quadratic system is an unsolved problem (except for some very simple cases). Therefore, following the approach proposed in [15], we will tackle the more practical problem of determining whether an assigned range of clinical conditions belongs to the region of attraction of the equilibrium point E_3. The approach proposed here can be used regardless of the system order, and it requires the solution of a particular type of LMI-based optimisation problem, namely the Generalised Eigenvalue Problem (GEVP) [40], for which efficient numerical optimisation routines exist [36]. In principle, if used with an appropriately validated model, such an approach could also be used to design an optimal and personalised strategy for cancer therapy.

3.6.2 Model of cancer development

In this section, we introduce the quadratic model of tumour growth developed in [39]. The model contains three state variables: the density of tumour cells M, the density of hunting predator cells N and the density of resting predator cells Z. The system model is

$$\begin{cases} \dot{M} = q + rM \left(1 - \frac{M}{k_1}\right) - \alpha MN \\ \dot{N} = \beta NZ - d_1 N \\ \dot{Z} = sZ \left(1 - \frac{Z}{k_2}\right) - \beta NZ - d_2 Z \end{cases} \tag{3.32}$$

where r is the growth rate of the tumour cells, q is the rate of conversion of the normal cells to the malignant ones, α is the rate of predation of the tumour cells by the hunting cells, β is the rate of conversion of the resting cells to the hunting cells, d_1 represents the natural death rate of the hunting cells, s is the growth rate of the resting predator cells, d_2 is the natural death rate of the resting cells, k_1 is the maximum carrying or packing capacity of the tumour cells and k_2 is the maximum carrying capacity of the resting cells. All these parameters are positive. In particular, for each cell population, k_1 and k_2 $(k_1 > k_2)$ represent the maximum number of cells that the environment could support in the absence of competition between these populations.

Note that in the model (3.32), the dynamics of the tumour–immune system interactions are described using a quadratic formulation. Indeed, for tumour cells and resting predator cells, the growth is modelled by adopting Lotka–Volterra and logistic terms [41]. In general, the logistic growth factor is defined as

$$R(x) = r \left(1 - \frac{x}{f}\right), \tag{3.33}$$

where x is the number of individuals of the population, and r and f are positive constants. For a given population, r is the intrinsic growth factor and f is the maximum number of individuals that can cohabit in the same environment such that each individual finds the necessary amount of resources for survival, denoted as the *carrying capacity*. From Eq. (3.33), note that $R(x)$ is a maximum when the population level is low, becomes zero when the population reaches the carrying capacity and is negative when this level is exceeded. Other papers in the literature, e.g. [42] and [43], also present ordinary differential equation models in which the tumour growth is described using logistic terms. Indeed, several recent clinical tests on measurable tumours have confirmed that, at higher tumour density, the growth of the tumour increases more slowly.

3.6.3　Stability of the equilibrium points

System (3.32) has three equilibrium points

$$E_1 = \left[\frac{k_1}{2} \left(1 + \sqrt{1 + \frac{4q}{rk_1}} \right), 0, 0 \right] \tag{3.34a}$$

$$E_2 = \left[\frac{k_1}{2} \left(1 + \sqrt{1 + \frac{4q}{rk_1}} \right), 0, k_2 \left(1 - \frac{d_2}{s} \right) \right] \tag{3.34b}$$

$$E_3 = \left[M^*, \frac{s}{\beta} \left(1 - \frac{d_1}{\beta k_2} \right) - \frac{d_2}{\beta}, \frac{d_1}{\beta} \right], \tag{3.34c}$$

where

$$M^* = \frac{ -\left[\frac{\alpha s}{\beta} \left(1 - \frac{d_1}{\beta k_2} \right) - \frac{\alpha d_2}{\beta} - r \right] + \sqrt{ \left[\frac{\alpha s}{\beta} \left(1 - \frac{d_1}{\beta k_2} \right) - \frac{\alpha d_2}{\beta} - r \right]^2 + \frac{4rq}{k_1} } }{ 2\frac{r}{k_1} }. \tag{3.35}$$

The three equilibrium points given above are biologically admissible only if they belong to the positive orthant (since concentrations of cells cannot take negative values). From Eq. (3.34a), we note that in the case of the equilibrium E_1 only malignant cells are present, and this equilibrium point always belongs to the positive orthant since the system parameters are positive. Both malignant cells and resting predator cells are present in the organism in the case of the equilibrium point E_2. Finally, when the system trajectories are around the equilibrium E_3, all three species of cells are present. To guarantee biological admissibility of the equilibria E_2 and E_3, it is necessary that $s > d_2$ in Eq. (3.34b) and

$$\beta > \frac{sd_1}{k_2(s - d_2)} \tag{3.36}$$

in Eq. (3.34c), respectively.

Regarding the stability properties of these equilibrium points, as shown in [39], the first equilibrium point is always unstable because the values of the system parameters are all positive. Also, if the equilibria E_1 and E_2 belong to the positive orthant and E_3 does not, E_2 is an asymptotically stable equilibrium point. Finally, if E_3 also belongs to the positive orthant, then E_2 is unstable and E_3 is asymptotically stable.

An asymptotically stable equilibrium point corresponds to a favourable condition from a clinical point of view, since it represents a dormant tumour in which the density of malignant cells does not vary.

Moreover, when E_3 belongs to the positive orthant, it is possible to decrease the steady-state density of the malignant cells by varying the rate of destruction of the tumour cells by the hunting cells (system parameter α). In addition, by comparing the density of the malignant cells in the equilibrium points E_3 and E_2, it is possible to verify that, if

$$\alpha < \frac{2r\beta}{s\left(1 - \frac{d_1}{\beta k_2}\right) - d_2}, \qquad (3.37)$$

the density of malignant cells in E_3 is lower than in E_2.

There is a biological interpretation for the equilibrium points E_2 and E_3 lying in the positive orthant. In particular, if only E_2 belongs to the positive orthant, a mechanism for converting resting predator cells to hunting predator cells does not exist. Conversely, when E_3 belongs to the positive orthant it is possible to control the steady-state density of the tumour by varying α. Therefore, it is of significant clinical interest to determine the region of attraction surrounding E_3, since this defines a *safety* region within which all state trajectories asymptotically return to the favourable clinical condition. As noted above, however, the exact computation of the region of attraction for all but the most simple quadratic systems is extremely difficult. Therefore, in the following we instead focus on the simpler problem of demonstrating that a specified region (in this case a box in the state-space) is included in the region of attraction of the equilibrium point. If this can be shown for a large enough box, then the goal of any therapeutic strategy should be to lead the system evolution from a given range of cell densities (corresponding to an initial point in the state-space) into such a box in the region of attraction of an asymptotically stable equilibrium in which the tumour cell density is null (tumour remission) or at least low (tumour dormancy) as in the case of E_3.

In [43] and [44], it has been shown that new protocols for cancer treatment, which make use of vaccines and immunotherapy, are able to control (or block) the tumour growth by modifying some critical parameters of the dynamical system that regulate the interactions between tumour cells and immune cells. Immunotherapy consists of the administration of therapeutic antibodies as drugs which can make immune cells able to kill more tumour cells. Therefore, in the following, we shall assume that we are able to control the immunotherapeutic action in the model (3.32), by varying the value of the parameter α. For the purposes of cancer therapy planning, an optimal value of α will be determined which is able to ensure the existence of a specified safety region around E_3.

3.6.4 Checking inclusion in the region of attraction

Let us consider a quadratic system in the form

$$\dot{x} = Ax + B(x), \qquad (3.38)$$

where $x \in \mathbb{R}^n$ is the system state and

$$
B(x) = \begin{pmatrix} x^T B_1 x \\ x^T B_2 x \\ \vdots \\ x^T B_n x \end{pmatrix}
\tag{3.39}
$$

with $B_i \in \mathbb{R}^{n \times n}$, $i = 1, \ldots, n$.

First note that the study of the stability properties of a nonzero equilibrium point of system (3.38) can always be reduced to the study of the corresponding properties of the origin of the state-space of another quadratic system by applying a suitable change of variable. Indeed assume that $x_e \neq 0$ is an equilibrium point for system (3.38); then

$$
Ax_e + B(x_e) = 0.
\tag{3.40}
$$

Now letting

$$
z = x - x_e,
\tag{3.41}
$$

it is readily seen that, from Eq. (3.40),

$$
\dot{z} = \left(A + 2 \begin{pmatrix} x_e^T B_1 \\ x_e^T B_2 \\ \vdots \\ x_e^T B_n \end{pmatrix} \right) z + B(z) + Ax_e + B(x_e)
$$

$$
= \left(A + 2 \begin{pmatrix} x_e^T B_1 \\ x_e^T B_2 \\ \vdots \\ x_e^T B_n \end{pmatrix} \right) z + B(z),
\tag{3.42}
$$

which is a quadratic system in the form of Eq. (3.38). Moreover, the equilibrium $z = 0$ of system (3.42) corresponds to the equilibrium $x = x_e$ of system (3.38).

On the basis of this observation, without loss of generality, we shall focus on the stability properties of the zero equilibrium point of system (3.38). Also, with a slight abuse of terminology, we shall refer to the "stability properties" of system (3.38), in place of the "stability properties of the zero equilibrium point" of system (3.38).

Checking local asymptotic stability of system (3.38) is rather simple, since it amounts to evaluating the eigenvalues of the linearised system $\dot{x} = Ax$. In the context of our analysis, however, establishing the local asymptotic stability is not enough, since it is required to check whether a given box around the equilibrium point in the state-space belongs to the region of attraction of the equilibrium.

In order to precisely define the problem, recall that a box $\mathcal{R} \subset \mathbb{R}^n$ can be described as follows:

$$\mathcal{R} = \operatorname{conv}\left\{x_{(1)}, x_{(2)}, \ldots, x_{(p)}\right\} \tag{3.43a}$$

$$= \left\{x \in \mathbb{R}^n : a_k^T x \le 1, \, k = 1, 2, \ldots, q\right\}, \tag{3.43b}$$

where p and q are suitable integer values, $x_{(i)}$ denotes the i-th vertex of $\mathcal{R}$ and $\operatorname{conv}\{\cdot\}$ denotes the operation of taking the convex hull of the argument.

For example, the box in $\mathbb{R}^2$

$$\mathcal{R} := [-1, 2] \times [-1, 3],$$

can be described both in the form of Eq. (3.43a) with

$$x_{(1)} = \begin{pmatrix} 2 & -1 \end{pmatrix}^T, \quad x_{(2)} = \begin{pmatrix} 2 & 3 \end{pmatrix}^T$$
$$x_{(3)} = \begin{pmatrix} -1 & 3 \end{pmatrix}^T, \quad x_{(4)} = \begin{pmatrix} -1 & -1 \end{pmatrix}^T,$$

and in the form of Eq. (3.43b) with

$$a_1^T = \begin{pmatrix} \frac{1}{2} & 0 \end{pmatrix}, \quad a_2^T = \begin{pmatrix} -1 & 0 \end{pmatrix}$$
$$a_3^T = \begin{pmatrix} 0 & \frac{1}{3} \end{pmatrix}, \quad a_4^T = \begin{pmatrix} 0 & -1 \end{pmatrix}.$$

In the next section we will try to solve the following problem:

Problem 1. Assume that the matrix A in Eq. (3.38) is Hurwitz (all eigenvalues of A have strictly negative real parts); then, given the box $\mathcal{R}$ defined in Eq. (3.43), such that 0 is an interior point of $\mathcal{R}$, establish whether $\mathcal{R}$ belongs to the region of attraction of system (3.38). $\diamond$

Let us first recall the following classical result from Lyapunov stability theory.

Estimate of the region of attraction: *A given closed set $E \subset \mathbb{R}^n$, such that 0 is an interior point of E, is an estimate of the region of attraction of system (3.38) if*

   i) E is an invariant set for system (3.38);

   ii) there exists a Lyapunov function $V(x)$ such that

      a) $V(x)$ is positive definite on E;

      b) $\dot{V}(x)$ is negative definite on E.

We choose a quadratic Lyapunov function $V(x) = x^T P x$, with P symmetric positive definite, so as to satisfy condition ii-a). The derivative of $V(x)$ along the trajectories of system (3.38) reads

$$\dot{V}(x) = \dot{x}^T P x + x^T P \dot{x}$$

$$= x^T \left\{ \left[A^T + (B_1^T x \ B_2^T x \ \dots \ B_n^T x) \right] P + P \left[A + \begin{pmatrix} x^T B_1 \\ x^T B_2 \\ \vdots \\ x^T B_n \end{pmatrix} \right] \right\} x < 0.$$

$$(3.44)$$

Note that the bracketed expression exhibits a linear dependence on the state variables $x_1, \dots, x_n$. This implies that it is negative definite on $\mathcal{R}$ if and only if it is negative definite *on the vertices* of $\mathcal{R}$. Hence, we can conclude that $V(x)$ satisfies condition ii) over $\mathcal{R}$ if the symmetric matrix function

$$A^T P + P A + P \begin{pmatrix} x^T B_1 \\ x^T B_2 \\ \vdots \\ x^T B_n \end{pmatrix} + \left(B_1^T x \ B_2^T x \ \cdots \ B_n^T x \right) P \qquad (3.45)$$

is negative definite on the vertices of $\mathcal{R}$. In order to also satisfy condition i), the idea is to enclose $\mathcal{R}$ into an invariant set which belongs to the region of attraction, namely the region bounded by a suitable level curve of the Lyapunov function. Based on the above ideas, and skipping the technical proof (the reader is referred to [15] for full details), Problem 1 can be transformed into the following Generalised Eigenvalue Problem (GEVP):

Problem 2. Find a scalar γ and a symmetric matrix P such that

$$0 < \gamma < 1$$
$$P > 0$$
$$\begin{pmatrix} 1 & \gamma a_k^T \\ \gamma a_k & P \end{pmatrix} \geq 0, \qquad k = 1, 2, \dots, 2n$$
$$x_{(i)}^T P x_{(i)} \leq 1, \qquad i = 1, 2, \dots, 2^n$$
$$\gamma (A^T P + P A) + P \left(B_1^T x_{(i)} \ B_2^T x_{(i)} \ \cdots \ B_n^T x_{(i)} \right)^T$$
$$+ \left(B_1^T x_{(i)} \ B_2^T x_{(i)} \ \cdots \ B_n^T x_{(i)} \right) P < 0, \qquad i = 1, 2, \dots, 2^n. \quad \Diamond$$

3.6.5 Analysis of the tumour dormancy equilibrium

Validation of the proposed technique

In what follows we use the model parameter values from [39]; thus $q = 10$, $r = 0.9$, $\alpha = 0.3$, $k_1 = 0.8$, $\beta = 0.1$, $d_1 = 0.02$, $s = 0.8$, $k_2 = 0.7$, $d_2 = 0.03$. Assume we want to establish whether the state response of system (3.32) converges to the asymptotically stable equilibrium $E_3 = \begin{bmatrix} 2.67 & 5.41 & 0.2 \end{bmatrix}^T$ after

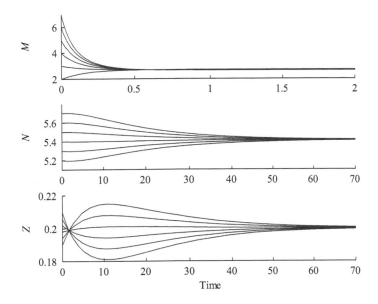

FIGURE 3.13: State response of system (3.32) from different perturbed initial conditions.

it has undergone a significant perturbation on the number of tumour cells. The convergence can be studied by simulation, as shown in Fig. 3.13, where the system evolution is computed for different initial conditions. However, this approach only allows us to test a finite number of initial points: to check the convergence over a whole region, we have to guarantee that it belongs to the domain of attraction of E_3, as follows.

In order to validate our technique, let us define a suitable box and solve Problem 1 for system (3.32). Define the box $\mathcal{R} = [2,7] \times [5.2, 5.7] \times [0.19, 0.21]$ containing the equilibrium point E_3 and the initial condition x_0. Since E_3 is a nonzero equilibrium point, we apply the change of variables (3.41) and study the properties of the zero state of the corresponding quadratic system in the form of Eq. (3.42) with

$$A = \begin{pmatrix} r & 0 & 0 \\ 0 & -d_1 & 0 \\ 0 & 0 & s-d_2 \end{pmatrix}, \quad B_1 = \begin{pmatrix} -\frac{r}{k_1} & -\frac{\alpha}{2} & 0 \\ -\frac{\alpha}{2} & 0 & 0 \\ 0 & 0 & 0 \end{pmatrix},$$

$$B_2 = \begin{pmatrix} 0 & 0 & 0 \\ 0 & 0 & \frac{\beta}{2} \\ 0 & \frac{\beta}{2} & 0 \end{pmatrix}, \quad B_3 = \begin{pmatrix} 0 & 0 & 0 \\ 0 & 0 & -\frac{\beta}{2} \\ 0 & -\frac{\beta}{2} & -\frac{s}{k_2} \end{pmatrix}. \tag{3.46}$$

Then we determine the vectors a_k for $k = 1, \ldots, 6$, the vertices $z_{(i)}$, for $i = 1, \ldots, 8$ and the expression for the box $\mathcal{R}$ translated to the origin. A solution to the feasibility Problem 2 is then given by

$$\gamma = 0.1194, \quad P = \begin{pmatrix} 0.0319 & -0.0027 & -0.0464 \\ -0.0027 & 4.194 & 7.259 \\ -0.0464 & 7.259 & 182.26 \end{pmatrix},$$

which can be readily obtained by numerically solving the problem using the YALMIP [35] package or the MATLAB® LMI Toolbox [36]. Thus, we can conclude that the box $\mathcal{R}$ belongs to the region of attraction of E_3. This implies that every trajectory starting from an initial condition included in $\mathcal{R}$ (such as, for example, those shown in Fig. 3.13) converges to E_3.

Optimisation of the therapeutic treatment

Changes in the hemodynamic perfusion of the tumour, radiation or chemotherapy may induce stochastic perturbations of the state variables around the equilibrium point E_3, thus leading the system away from the steady-state condition. If these perturbations were to bring the system out of the domain of attraction of the equilibrium point E_3, the state trajectories could diverge, leading to unbounded growth of the tumour. Given the above, the results presented in the following are potentially useful not only for the analysis of tumour development, but also to devise an effective therapeutic strategy: given a certain operative range, a suitable strategy could be that of enforcing the system trajectories to tend to the asymptotically stable equilibrium E_3 (tumour growth blocked). This problem can be translated in mathematical terms to that of computing the value of certain parameters such that the region of attraction of E_3 contains the given operative range.

By using the results presented in Section 3.6.4, it is possible to optimise the value of α, which depends on the amount of immunotherapeutic drug injected into the patient, in order to guarantee a safety region (i.e. included in the region of attraction) around the equilibrium point E_3. Thus, the box $\mathcal{R}$ is assigned in terms of an admissible variation interval for each state variable around the equilibrium point. The sizes of such intervals can be chosen on the basis of clinical knowledge about the admissible perturbations acting on the system.

In order to exemplify this strategy, we shall apply the proposed methodology to the quadratic model (3.32) using the same parameter values given in [39], except for α, which will be optimised as described below. First of all, note that the first component of the equilibrium E_3 depends on α, while the other components do not:

$$E_3(\alpha) = \begin{pmatrix} x(\alpha) & 5.41 & 0.2 \end{pmatrix}^T, \tag{3.47}$$

where $x(\alpha)$ is given by Eq. (3.35) when all parameters, except α, assume the values given in [39]. The first component of E_3 monotonically varies from 3.4 for $\alpha = 0$ (i.e. no therapy) to 2.68 for $\alpha = 0.3$ (maximum value of α compatible with Eq. (3.37)). In [39] the maximum allowable value of α is considered to simulate the system behaviour. Here our goal is to use the proposed approach to guarantee a reasonable safety region around E_3 while, at the same time, minimising the value of α and therefore the amount of drugs that need to be delivered to the patient. To cope with relatively large variations of the cell densities, we will take the box $\mathcal{R}$ defined in the previous section as the *safety* operating region for the system under investigation.

Thus, for a given value of α, we can

1. Compute the corresponding equilibrium point $E_3(\alpha)$ by using Eq. (3.47);

2. Determine whether $\mathcal{R}$ belongs to the region of attraction of $E_3(\alpha)$ by using the approach of Section 3.6.4.

By repeating these two steps for decreasing values of α, it is possible to find the minimum value which guarantees the existence of the specified safety region. In our case the result is $\alpha_{opt} = 0.08$, with $E_3(\alpha_{opt}) = \begin{bmatrix} 3.20 & 5.41 & 0.2 \end{bmatrix}^T$. Indeed it is possible to verify that, with the values given above,

$$\gamma = 0.203, \quad P = \begin{pmatrix} 0.0298 & -0.0227 & -0.3129 \\ -0.0227 & 5.862 & 18.37 \\ -0.3129 & 18.37 & 488.08 \end{pmatrix} \quad (3.48)$$

is an admissible solution to Problem 2.

In Fig. 3.14 the box R, the ellipsoidal invariant set determined by the Lyapunov function $x^T P x \leq 1$, with P given in Eq. (3.48), and several trajectories starting from different points in $\mathcal{R}$ are depicted. As expected, all trajectories which start from points in the safety box converge to the tumour dormancy equilibrium.

It is interesting to note that the trajectories can exit the box, exhibiting an overshoot which extends well outside the box, before reaching the equilibrium. Indeed, the fact that the initial condition belongs to the region of attraction does not guarantee that the number of malignant cells is bounded during the transient, rather it ensures that, after a possible overshoot, the system will return to the dormancy level. On the other hand, the proposed analysis method also provides a bound on the admissible system trajectories, which is given by the ellipsoidal region surrounding the box.

Developing an improved understanding of the dynamical behavior of tumour progression can have interesting implications for the development of therapeutic strategies. For example, a quantitative analysis of tumour growth over a time interval, coupled with an effective model, could help to determine whether the current therapy is effective and the observed growth is just a transient phenomenon, or the system has left the safety region and has entered a phase of unbounded growth requiring a different therapeutic action. Also note that the optimal value of α found in our analysis is very small. Under the assumption that the parameter α is representative of the dose of drug administered to the patient, the small value of α required in our calculations is an

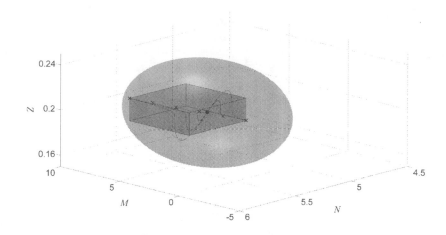

FIGURE 3.14: Polytope $\mathcal{R}$, ellipsoidal invariant set surrounding $\mathcal{R}$ and several trajectories starting from different points (cross markers) and converging to the tumour dormancy equilibrium (circle marker)

alluring result, because it suggests that, by exploiting the proposed technique, it is possible not only to devise a robust therapeutic strategy but also to minimise, at the same time, the amount of drug delivered to the patient. A major remaining challenge in immunotherapy, in fact, consists of improving antitumour activity without inducing unmanageable toxicity to normal tissues [45]. Therefore, a primary goal of current research in this area is to determine the minimum dose of drug capable of producing the desired effective therapeutic action, in order to limit unwanted side effects on normal tissues.

3.7 Case Study IV: Global optimisation of a model of the tryptophan control system against multiple experiment data

Biology background: *Tryptophan* is one of the essential amino acids (protein building-blocks) in humans, i.e. it cannot be synthesised internally and must be part of our diet. Tryptophan is also a protein precursor for serotonin and melatonin. A protein precursor is an inactive protein (or peptide) that can be turned into an active form by post-translational modification. Protein precursors are often used by an organism when the subsequent protein is potentially harmful, but needs to be available at short notice and/or in large quantities. Serotonin is a neurotransmitter that performs numerous functions in the human body including the control of appetite, sleep, memory and learning, temperature regulation, mood, behaviour, cardiovascular function, muscle contraction, endocrine regulation and depression. Melatonin is an important hormone that plays a role in regulating the circadian sleep-wake cycle. It also controls essential functions such as metabolism, sex drive, reproduction, appetite, balance, muscular coordination and the immune system response in fighting off various diseases.

Tryptophan has been shown to be effective as a sleep aid and antidepressant, and has been indicated for a range of other potential therapeutic applications including relief of chronic pain and the reduction of various impulsive, manic and violent disorders. It is sold as a prescription drug and is also available as a dietary supplement.

The biotechnology industry uses fermentation processes to commercially produce tryptophan. Large quantities of wild-type or genetically modified bacteria are grown in vats, and the food supplement is extracted from the bacteria and purified. Unfortunately, however, yields of tryptophan generated via this process are significantly lower than those achieved in microbial production of other amino acids, making its production an expensive and challenging process. This is almost certainly due to the exquisitely complex control system employed by the cell to regulate tryptophan production.

As has been pointed out by numerous researchers working in this area, the development of an improved quantitative understanding of this complex dynamical system is the obvious starting point in developing yield optimisation strategies for industrial tryptophan production.

3.7.1   Introduction

Many cellular control systems employ multiple feedback loops to allow fast
and efficient adaptation to uncertain environments. The various feedback
mechanisms used by prokaryotes such as *E. coli* to regulate the expression
of proteins involved in the production of the amino acid tryptophan combine
to form an extremely complex, but highly effective, feedback control system.
This system has been the subject of numerous modelling studies in recent
systems biology research, with the result that a plethora of different mathematical models of tryptophan regulation may now be found in the literature;
see, for example, [46]–[49] and references therein. In each of these modelling
studies, the dynamics of the proposed model were compared with an extremely
limited set of experimental data, and our current understanding of the underlying reactions is such that very little information is available to guide the
selection of parameter values for the models. As a result, in most previous
studies only qualitative agreement between model outputs and experimental
data could be demonstrated; see, for example, [46]. Since many of the models
in the literature have been derived using diverse assumptions about the exact workings of the underlying feedback mechanisms involved, the lack of any
strong validation (or invalidation) of a particular model has hindered progress
in understanding the underlying design principles of this system.

   In this Case Study, we focus on the issue of model validation and proceed
from the assumption that, for a valid model, there must exist at least one
set of biologically plausible model parameters which yields a close match to
the available experimental data. We consider one particular model of the
tryptophan control system introduced in [46], which includes regulation of
the trp operon by feedback loops representing repression, feedback inhibition
and transcriptional attenuation [46]. The model also incorporates the effect
of tryptophan transport from the growth medium as well as the various time
delays involved in the transcription and translation processes. We use global
optimisation methods to investigate whether, for the proposed model structure, realistic (i.e. biologically plausible) parameter values can be found so
that the model reproduces the dynamic response of the *in vitro* system to a
number of different experimental conditions, [50].

3.7.2   Model of the tryptophan control system

The mathematical model of the tryptophan control system considered in this
study is taken from [46] and consists of the set of nonlinear differential equations (3.49).

$$\frac{dO_F(t)}{dt} = \frac{K_r}{K_r + \dfrac{T(t)}{T(t) + K_t}R} \left\{ \mu O - k_p P \left[O_F(t) - O_F(t - \tau_p)e^{-\mu \tau_p} \right] \right\} - \mu O_F(t)$$

$$(3.49a)$$

$$\frac{dM_F(t)}{dt} = k_p P O_F(t - \tau_m) e^{-\mu \tau_m} \left[1 - b \left(1 - e^{T(t)/c} \right) \right] \tag{3.49b}$$

$$- k_\rho \rho \left[M_F(t) - M_F(t - \tau_\rho) e^{-\mu \tau_\rho} \right] - (k_d D + \mu) M_F(t) \tag{3.49c}$$

$$\frac{dE(t)}{dt} = \frac{1}{2} k_\rho \rho M_F(t - \tau_e) e^{-\mu \tau_e} - (\gamma + \mu) E(t) \tag{3.49d}$$

$$\frac{dT(t)}{dt} = K \frac{K_i^{n_H}}{K_i^{n_H} + T^{n_H}(t)} E(t) - g \frac{T(t)}{T(t) + K_g} + d \frac{T_{\text{ext}}}{e + T_{\text{ext}} \left[1 + T(t)/f \right]} - \mu T(t) \tag{3.49e}$$

In Eq. 3.49, R is total repressor concentration, O is total operon concentration, P is mRNA polymerase (mRNAP) concentration, $O_F(t)$ is free operon concentration, $M_F(t)$ is free mRNA concentration, $E(t)$ is total enzyme concentration, $T(t)$ is tryptophan concentration, K_r is the repression equilibrium constant, K_t is the the rate equilibrium constant between the total repressor and the active repressor, μ is the growth rate, k_p is the DNA-mRNAP isomerisation rate, b and c are constants defining the dynamics of the transcriptional attenuation, k_ρ is the mRNA-ribosome isomerisation rate, ρ is the ribosomal concentration, k_d is the mRNA degradation rate, D is the mRNA degrading enzyme, γ is the enzymatic degradation rate constant, K is the tryptophan production rate, which is proportional to the active enzyme concentration, K_i is the equilibrium constant for the *Trp* feedback inhibition of anthranilate synthase reaction, which is modelled by a Hill equation with the coefficient n_H and g is the maximum tryptophan consumption rate.

The internal tryptophan consumption is modelled by a Michaelis–Menten type term with the constant K_g; T_{ext} is the external tryptophan uptake, d, e and f are parameters describing the dynamics of the external tryptophan uptake rate, τ_p is the time taken for mRNAP to bind to DNA and move away to free the operon, τ_m is the time taken for mRNA to be produced after mRNAP binds to the DNA, τ_ρ is the time taken for the ribosome to bind to mRNA and initiate translation and τ_e is another ribosome binding rate delay for the enzyme.

All 25 independent parameters are given in Table 3.4 and the dependent parameters are calculated as follows: $\bar{T} = K_i$, $k_\rho = 1/(\rho \tau_\rho)$, $kdD = \rho k_\rho/30$, $K_g = \bar{T}/20$, $g = T_{\text{cr}}(\bar{T} + K_g)/\bar{T}$, $\bar{E}_A = \bar{E} K_i^{n_H}/(K_i^{n_H} + \bar{T}^{n_H})$, $\bar{G} = g\bar{T}/(\bar{T} + K_g)$ and $K = (\bar{G} + \mu\bar{T})/\bar{E}_A$, where $\bar{T}$ and $\bar{E}$ are the steady-state of tryptophan and enzyme, respectively, and T_{cr} is the tryptophan consumption rate. More detail about the model can be found in [46]. The model given above clearly takes into account the three different feedback control mechanisms (repression, feedback inhibition and transcriptional attenuation) that have been experimentally verified to operate in the tryptophan operon. In [46], the authors were also careful to base their choices for model parameters on the available biological data, although in many cases little information is available.

For validation purposes, experimental data is available in [50], which reports the results of a number of experiments with wild and mutant strains of the *E. coli* CY15000 strain. These experiments consisted of growing bacteria in a number of different media which included tryptophan until the culture

Table 3.4: Original and optimised parameters in the tryptophan model

	Unit	Original in [46]	Optimal Exp. A	Optimal Exp. B	Optimal Exp. C
d	$[\cdot]$	23.5	23.5	23.5	23.5
e	$[\cdot]$	0.9	0.9	0.9	0.9
f	$[\cdot]$	380	380	380	380
R	$[\mu M]$	0.8	1.357	1.759	1.518
O	$[\mu M]$	0.0033	0.0059	0.0086	0.0039
P	$[\mu M]$	2.6	3.22	2.86	3.33
$\bar{E}$	$[\mu M]$	0.378	0.338	0.550	0.349
T_{cr}	$[\mu M/min]$	22.7	14.07	14.00	14.01
K_r	$[\mu M]$	0.0026	0.0015	0.0022	0.0077
K_t	$[\mu M]$	60.34	64.55	158.84	127.41
K_i	$[\mu M]$	4.09	6.93	53.13	50.41
n_H	$[\cdot]$	1.2	1.00	1.00	1.00
b	$[\cdot]$	0.85	0.53	0.33	0.60
c	$[\cdot]$	0.04	0.0083	0.345	0.0109
ρ	$[\mu M]$	2.9	3.33	3.68	4.00
γ	$[1/min]$	0.0	0.0113	0.00063	0.0034
μ	$[1/min]$	0.01	0.0264	0.0245	0.0192
τ_p	$[min]$	0.1	0.0267	0.0381	0.0664
τ_m	$[min]$	0.1	0.0277	0.2587	0.2241
τ_ρ	$[min]$	0.05	0.0874	0.1300	0.1299
τ_e	$[min]$	0.66	1.1284	1.7131	1.7156
$O_F(0)$	$[\mu M]$	$4.8765e-5$	$7.6444e-5$	$0.9772e-5$	$0.9965e-5$
$M_F(0)$	$[\mu M]$	$1.2037e-4$	$0.4304e-4$	$0.3677e-4$	$0.2667e-4$
$E(0)$	$[\mu M]$	0.0119	0.0238	0.0238	0.0238
$T(0)$	$[\mu M]$	16.571	13.962	42.772	42.967

reached a steady-state. Then the bacteria were washed and put into the same media without tryptophan. The response of enzyme anthranilate synthase to these nutritional shifts was then measured as a function of time. Anthranilate synthase is the key enzyme involved in tryptophan biosynthesis and its activity is proportional to the production rate of tryptophan. In Fig. 3.15, the dash-dot lines give the model responses according to the three different experimental setups. As shown in the figure, the steady-state values are close to the experimental data but there are large discrepancies in the transient dynamics. This leaves open the question: is this discrepancy simply a result of an incorrect choice of parameters in the model, or are the underlying assumptions on which the model is constructed (its structure) a poor representation of the biological reality? In the next section, we show how global optimisation

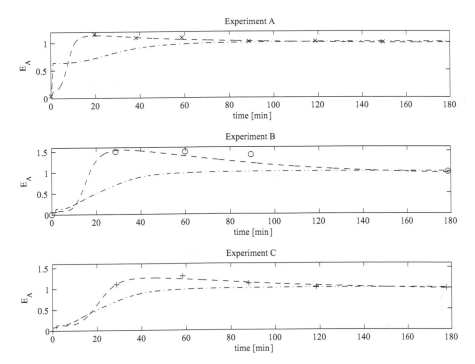

FIGURE 3.15: Optimised (dashed line) versus original (dash-dot line) model responses for data from experiment A (x), experiment B (o) and experiment C (+); experimental data are taken from [50].

can be used to at least partially resolve this issue.

3.7.3 Model analysis using global optimisation

For each set of experimental data, we formulate an optimisation problem to minimise the square sum of errors between the dynamics of the active enzyme concentration produced by the model and the experimental data as follows:

$$\min_{p} J = \sum_{j=1}^{N} |\bar{x}(t_j) - \tilde{x}(t_j)|^2 \qquad (3.50)$$

where p is the set of parameters in the model, $\tilde{x}(t_j)$ is the experimental measurement at time t_j and $\bar{x}(t_j)$ is the model response at time t_j. This nonlinear and non-convex optimisation problem is solved using a hybrid Genetic Algorithm based on the one developed in [51] for each of the three different sets of experimental data. The results are shown in Fig. 3.15 and Table 3.4. As can be seen from the figure, while the original model shows quite a poor agreement with the data, the optimised model is able to almost exactly reproduce the responses of the *in vitro* system for each different experiment. Importantly, the optimal model parameters are also all within biologically plausible ranges. Although these results are obviously very far from being a comprehensive validation of the proposed model, they do show that the proposed model structure *can* accurately reproduce the experimentally measured behaviour, hence making the assumptions on which the model is based a plausible explanation of the biological processes involved.

References

[1] Gyuk I. Signal transduction and nonlinearity. In *Proceedings of the EMF Engineering Review Symposium*, Charleston, South Carolina, USA, April 28-29, 1998.

[2] Keener J and Sneyd J. *Mathematical Physiology*. Berlin: Springer-Verlag, 2nd edition, 2009.

[3] d'Alche-Buc F and Brunel N. Estimation of parametric nonlinear ODEs for biological networks identification. In Lawrence N, Girolami M, Rattray M and Sanguinetti G, editors, *Learning and Inference in Computational Systems Biology*, Boston: MIT Press, 2009.

[4] Elowitz MB and Leibler S. A synthetic oscillatory network of transcriptional regulators. *Nature*, 403:335–338, 2000.

[5] de Pillis LG, Radunskaya AE, and Wiseman CL. A validated mathematical model of cell-mediated immune response to tumor growth. *Cancer Research*, 65:(17), 2005.

[6] Khalil HK. *Nonlinear Systems*. Upper Saddle River: Prentice Hall, 2002.

[7] Nelson DL and Cox MM. *Lehninger Principles of Biochemistry*. W.H. Freeman and Co., 3rd edition, 2000.

[8] Goldbeter A, Dupont G, and Berridge MJ. Minimal model for signal–induced Ca^{2+} oscillations and for their frequency encoding through protein phosphorylation. *PNAS*, 87:1461–1465, 1990.

[9] Banga JR. Optimization in computational systems biology. *BMC Systems Biology*, 2(47), 2009.

[10] Zwolak JW, Tyson JJ, and Watson LT. Globally optimised parameters for a model of mitotic control in frog egg extracts. *IET Systems Biology*, 152(2):81–92, 2005.

[11] Chen WW, Schoeberl B, Jasper PJ, Niepel M, Nielsen UB, Lauffen-burger DA, and Sorger PK. Input–output behavior of ErbB signaling pathways as revealed by a mass action model trained against dynamic data. *Molecular Systems Biology*, 5:239, 2009.

[12] Cosentino C, Curatola W, Montefusco F, Bansal M, di Bernardo D, and Amato F. Linear matrix inequalities approach to reconstruction of biological networks. *IET Systems Biology*, 1(3):164–173, 2007.

[13] Vilela M, Chou I-C, Vinga S, Vasconcelos ATR, Voit EO, and Almeida JS. Parameter optimization in s-system models. *BMC Systems Biology*, 2:35, 2008.

[14] Kim J, Bates DG, Postlethwaite I, Ma L, and Iglesias P. Robust-ness analysis of biochemical network models. *IET Systems Biology*, 152(3):96–104, 2006.

[15] Merola A, Cosentino C, and Amato F. An insight into tumor dormancy equilibrium via the analysis of its domain of attraction. *Biomedical Signal Processing and Control*, 3:212–219, 2008.

[16] Maryn-Sanguino A and Torres NV. Optimization of tryptophan pro-duction in bacteria: design of a strategy for genetic manipulation of the tryptophan operon for tryptophan flux maximization. *Biotechnology Progress*, 16(2):133-145, 2000.

[17] Torres NV and Voit EO. *Pathway Analysis and Optimization in Metabolic Engineering*. Boston: Cambridge University Press, 2002.

[18] Sun C-P, Usui T, Yu F, Al-Shyoukh I, Shamma J, Sun R, and Ho C-M. Integrative systems control approach for reactivating Kaposi's sarcoma-associated herpesvirus (KSHV) with combinatory drugs. *Integrative Biology*, 1:123–130, 2009.

[19] Dasika MS and Maranas CD. OptCircuit: An optimization based method for computational design of genetic circuits. *BMC Systems Biology*, 2(24), 2008.

[20] Lebiedz D. Exploiting optimal control for target-oriented manipulation of (bio)chemical systems: A model-based approach to specific modi-fication of self-organized dynamics. *International Journal of Modern Physics B*, 19(25):3763–3798, 2005.

[21] Balsa-Canto E, Alonso AA, and Banga JR An optimal identification procedure for model development in systems biology. In *Proceedings of the FOSBE (Foundations of Systems Biology and Engineering) Confer-ence*, Stuttgart, Germany, 2007.

[22] *Optimization Toolbox User's Guide, Version 2*. The MathWorks, September 2000.

[23] Goldberg DE. *Genetic Algorithms in Search, Optimization and Machine Learning.* Boston: Addison-Wesley, 1989.

[24] Fleming PJ and Purshouse RC. Evolutionary algorithms in control systems engineering: a survey. *Control Engineering Practice,* 10:1223–1241, 2002.

[25] Storn R and Price K. Differential evolution: a simple and efficient heuristic for global optimization over continuous space. *Journal of Global Optimization,* 11:341–369, 1997.

[26] Lampinen J and Zelinka I. Mechanical engineering design by differential evolution. In Corne D, Dorigo M, and Glover F (eds.) *New Ideas in Optimisation,* McGraw-Hill, London (UK), 127–146, 1999.

[27] Rogalsky T, Derksen RW, and Kocabiyik S. Differential evolution in aerodynamic optimization. *Canadian Aeronautics and Space Institute Journal,* 46:183–190, 2000.

[28] Jones DR, Perttunen CD, and Stuckman BE. Lipschitzian optimization without the Lipschitz constant. *Journal of Optimization Theory and Application,* 9, 1993.

[29] Finkel DE and Kelley CT. Convergence analysis of the DIRECT algorithm. *N.C. State University Center for Research in Scientific Computation Tech. Report number CRSC-TR04-28,* 2004.

[30] Zhu H and Bogy DB. DIRECT algorithm and its application to slider air-bearing surface optimisation. *IEEE Transactions on Magnetics,* 5, 2002.

[31] Carter R, Gablonsky JM, Patrick A, Kelley CT, and Eslinger OJ. Algorithms for noisy problems in gas transmission pipeline optimization. *Optimization and Engineering,* 2:139–157, 2001.

[32] Yen J, Liao JC, Randolph D, and Lee B. A hybrid approach to modeling metabolic systems using genetic algorithm and simplex method. In *Proceedings of the 11th IEEE Conference on Artificial Intelligence for Applications,* 277–283, Los Angeles, CA, Feburary 1995.

[33] Lobo FG and Goldberg DE. Decision making in a hybrid genetic algorithm. *IlliGAL Report No. 96009,* September 1996.

[34] *Genetic Algorithm and Direct Search Toolbox User's Guide, Version 2,* The MathWorks, New York, September 2005.

[35] Lofberg J. YALMIP: Software for solving convex (and nonconvex) optimization problems. In *Proceedings of the American Control Conference, Minneapolis,* 2006.

[36] Gahinet P, Nemirovski A, Laub AJ, and Chilali M. *LMI Control Toolbox.* The Mathworks Inc, Natick, 1995.

[37] Adam J and Bellomo N. *A Survey of Models on Tumor Immune Systems Dynamics*. Birkhauser, Boston, 1996.

[38] Preziosi L. From population dynamics to modelling the competition between tumors and immune system. *Mathematical and Computational Modelling*, 23(6):132–152, 2003.

[39] Sarkar RR and Banerjee S. Cancer self remission and tumor stability — a stochastic approach. *Mathematical Biosciences*, 196(1):65–81, 2005.

[40] Boyd S, El Ghaoui L, Feron E, and Balakrishnan V. *Linear Matrix Inequalities in System and Control Theory*. Philadelphia: SIAM Press, 1994.

[41] Murray JD. *Mathematical Biology*. Berlin Heidelberg: Springer–Verlag, New York, 2002.

[42] Kuznetsov VA, Maklkin IA, Taylor MA, and Perelson AS. Nonlinear dynamics of immunogenic tumors: parameter estimation and global bifurcation analysis. *Bullettin of Mathematical Biology*, 56(2):295–321, 1994.

[43] de Pillis LG, Gu W, and Radunskaya AE. Mixed immunotherapy and chemotherapy of tumors: modeling applications and biological interpretations. *Journal of Theoretical Biology*, 238(4):841–862, 2006.

[44] Gatenby RA and Vincent TL Application of quantitative models from population biology and evolutionary game theory to tumor therapeutic strategies. *Molecular Cancer Therapy*, 2(2):919–927, 2003.

[45] Blattman JN and Greenberg PD. Cancer immunotherapy: a treatment for the masses. *Science*, 305:200–205, 2004.

[46] Santillan M and Mackey MC. Dynamic regulation of the tryptophan operon: a modeling study and comparison with experimental data. *PNAS*, 98(4):1364–1369, 2001.

[47] Xiu Z-L, Chang ZY, and Zeng A-P. Nonlinear dynamics of regulation of bacterial trp operon: Model analysis of integrated effects of repression, feedback inhibition, and attenuation. *Biotechnology Progress*, 18(4):686-693, 2002.

[48] Bhartiya S, Rawool S, and Venkatesh KV. Dynamic model of Escherichia coli tryptophan operon shows an optimal structural design. *Journal of European Biochemistry*, 270(12):2644-2651, 2003.

[49] Santillan M and Zeron ES. Analytical study of the multiplicity of regulatory mechanisms in the tryptophan operon. *Bulletin of Mathematical Biology*, 68(2):343-359, 2006.

[50] Yanofsky C and Horn V. Role of regulatory features of the trp operon of Escherichia coli in mediating a response to a nutritional shift. *Journal of Bacteriology*, 176(20):6245-6254, 1994.

[51] Menon PP, Kim J, Bates DG, and Postlethwaite I, Clearance of nonlinear flight control laws using hybrid evolutionary optimisation. *IEEE Transactions on Evolutionary Computation*, 10(6):689-699, 2006.

4

Negative feedback systems

4.1 Introduction

Negative feedback is a powerful mechanism for changing and controlling the dynamics of a system. Through the expert use of this type of feedback, control engineers are able to manipulate the dynamics of a huge variety of different systems, so that they behave in a way that is desirable and efficient from the point of view of the user, [1, 2, 3]. In biological systems, evolutionary pressures have led to the use of negative feedback for a wide variety of purposes, including homeostasis, chemotaxis, adaptation and signal transduction. As shown in Fig. 4.1, the principle of negative feedback is extremely simple: a feedback loop is closed around a system G and the measured output of the system y is compared to its desired value r. The resulting error signal e is acted on by a controller K, which generates an input signal u for the system which causes its output to move towards its desired value. Note that, depending on the type of system, and the level of control required, the controller K could be as simple as a unity gain or as complex as a high-order nonlinear dynamical system. Consider, for example, a simple first-order system $G(s)$

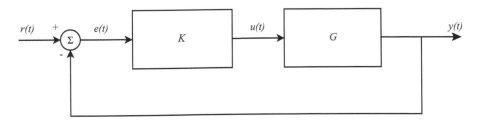

FIGURE 4.1: Negative feedback control scheme.

which has a time constant of 3 seconds and a system gain of 10:

$$Y(s) = G(s)U(s); \quad G(s) = \frac{10}{3s + 1} \tag{4.1}$$

The response of this system to a step input $U(s) = 1/s$ is shown in Fig. 4.2, and as expected, the system takes 3 seconds to reach 63% of its final value. Suppose the response of the system is now required to be much faster than

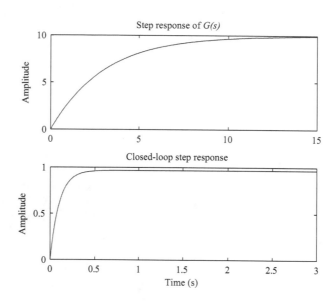

FIGURE 4.2: Step responses of $G(s) = 10/(3s+1)$ with and without feedback control.

this — the time constant can be changed by placing a simple static controller with gain K in series with the system and "closing the loop" using negative feedback, as shown in Fig. 4.1. The open-loop transfer function (i.e. the transfer function from $R(s)$ to $Y(s)$ without any feedback) for the system is now given by $L(s) = KG(s) = 10K/(3s + 1)$ while the closed-loop transfer function from $R(s)$ to $Y(s)$ is given by

$$T(s) = \frac{L(s)}{1 + L(s)} = \frac{10K}{3s + 1 + 10K} \tag{4.2}$$

The time constant of the system has now changed and is dependent on the value of K. To see this, we divide by $1 + 10K$ to get a unit constant coefficient on the denominator:

$$T(s) = \frac{\frac{10K}{1+10K}}{\frac{3}{1+10K}s + 1} \tag{4.3}$$

Thus the gain and time constant of the closed-loop system are now given by $\frac{10K}{1+10K}$ and $\frac{3}{1+10K}$, respectively. Incorporating even a modest value of, say,

2.9 for the gain K thus results in a dramatically faster response of the system, which now has a time constant of 0.1 seconds and responds to a step change in R, as shown in Fig. 4.2.

At this point the reader might be tempted to ask: if the aim is to produce a required type of dynamic response, why not just place a controller in series with the system to achieve that response, rather than going to the trouble of using feedback? Indeed, if the dynamics of the system were precisely known (and not subject to any variation), and the system operated in a vacuum containing no external disturbances, then there would be no need to use feedback control. This is never the case, however, and as we shall see later in this chapter, it is the ability of feedback to generate insensitivity ("robustness") in the response of systems to variations and disturbances that leads engineers (and bacteria) to use it.

Are there any limitations to the type of dynamics that can be imposed on a particular system by exploiting the power of feedback? The answer, of course, is that there are, and indeed the study of these fundamental limitations has kept control theorists busy for many decades. In the case of the system above, for example, we can see that while both the steady-state gain and the time constant of the closed-loop system can be set by choosing an appropriate value for the controller gain K, these two characteristics cannot be adjusted *independently* (at least using this type of simple controller). As another example, consider the system

$$G(s) = \frac{100}{s^2 + 8s + 10} \tag{4.4}$$

This system has the step response shown in Fig. 4.3, with a steady-state gain of 10 and a rise time of 1.5 seconds. Suppose now that it is required to lower the gain of this system so that it no longer amplifies input signals (gain of 1) and that we again require a significantly faster response. This can be achieved by placing a simple static controller K, this time with a gain of 20, in a negative feedback loop around the system, as shown in Fig. 4.1. The resulting closed-loop step response, shown in Fig. 4.3, delivers unity steady-state gain with a much faster rise time of 0.02 seconds; however, the response is now also much more oscillatory, with a significant initial overshoot. Worse still, if even a very small time delay of 5 milliseconds is included in the feedback loop, as shown in Fig. 4.4, the closed-loop response of the system actually becomes unstable (Fig. 4.3). The above example illustrates a fundamental point about the use of feedback: its power to radically change (and even destabilise) a system's dynamics makes it a potentially dangerous strategy for achieving control. The precise way in which biological systems have evolved to guard against the potentially dangerous effects of feedback is only just starting to be elucidated in recent systems biology research. Potentially of even more importance for medical applications is to understand how such safeguards sometimes fail, since the resulting unstable behaviour is postulated to be at the root cause of many diseases, e.g. uncontrolled growth of tumour cells.

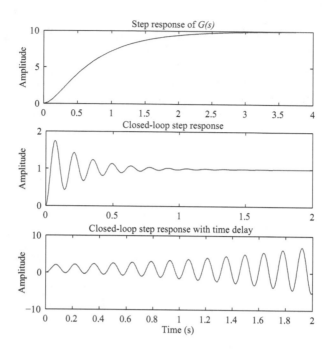

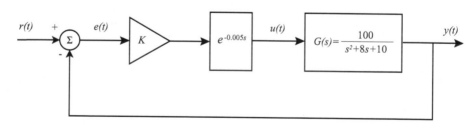

FIGURE 4.3: Step responses of $G(s) = 100/(s^2+8s+10)$, $G(s)$ under feedback control and $G(s)$ under feedback control with time delay.

FIGURE 4.4: Negative feedback control scheme with time delay.

In later sections of this chapter we will return to explore the potential uses and fundamental limitations of negative feedback in more detail. First, however, we provide some basic tools for evaluating the stability of feedback loops. In order to make the exposition as clear as possible we will focus on linear systems, discussing limitations and extensions of the results to the case of nonlinear systems as appropriate.

4.2 Stability of negative feedback systems

In this section, we introduce an important tool for determining the stability of linear feedback systems — Nyquist's Stability Criterion. In contrast to the tests for stability described in previous chapters, this criterion allows us to gain meaningful insight into the *degree* of stability of a feedback control system, and paves the way for the introduction of notions of robustness which will be further developed in later chapters. Nyquist's Stability Criterion is based on a result from complex analysis known as Cauchy's principle of the argument, which may be stated as follows:

Let $F(s)$ be a function which is differentiable in a closed region of the complex plane s except at a finite number of points (namely, the poles of $F(s)$). Assume also that $F(s)$ is differentiable at every point on the contour of the region. Then, as s travels around the contour in the s-plane in the clockwise direction, the function $F(s)$ encircles the origin in the $(\mathrm{Re}\{F(s)\},\mathrm{Im}\{F(s)\})$-plane in the same direction N times, where $N = Z - P$ and Z and P are the number of zeros and poles (including their multiplicities) of the function inside the contour.

The above result can be also written as $\arg\{F(s)\} = (Z - P)2\pi = 2\pi N$, which explains the term "principle of the argument."

Now consider a closed-loop negative feedback system

$$T(s) = \frac{G(s)}{1 + G(s)K(s)}$$

where $G(s)$ represents the system and $K(s)$ is the feedback controller. Since the poles of a linear system are given by those values of s at which its transfer function is equal to infinity, it follows that the poles of the closed-loop system may be obtained by solving the following equation:

$$1 + G(s)K(s) = D(s) = 0$$

This equation is known as the *characteristic equation* for the closed-loop system. Thus the zeros of the complex function $D(s)$ are the poles of the closed-loop transfer function. In addition, it is easy to see that the poles of $D(s)$ are the zeros of the closed-loop system $T(s)$. Nyquist's Stability Criterion is obtained by applying Cauchy's principle of the argument to the complex function $D(s)$, as follows.

The Nyquist plot is a polar plot of the function $D(s)$ as s travels around the contour given in Fig. 4.5. Note that the contour in this figure covers the whole unstable half (right-hand side) of the complex plane s, in the limit as $R \to \infty$. Since the function $D(s)$, according to Cauchy's principle of the argument, must be analytic at every point on the contour, any poles of $D(s)$ on the imaginary axis must be encircled by infinitesimally small semicircles, as shown in Fig. 4.5. We are now ready to state the Nyquist Stability Criterion:

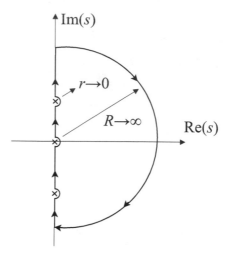

FIGURE 4.5: Nyquist contour in the s-plane.

The number of unstable closed-loop poles is equal to the number of unstable open-loop poles plus the number of encirclements (counted as positive in the clockwise and negative in the counter-clockwise direction) of the origin by the Nyquist plot of $D(s)$.

This result follows directly by applying Cauchy's principle of the argument to the function $D(s)$ with the s-plane contour given in Fig. 4.5, and noting that

1. Z and P represent the numbers of zeros and poles, respectively, of $D(s)$ in the right half plane, and

2. the zeros of $D(s)$ are the closed-loop system poles, while the poles of $D(s)$ are the open-loop system poles (closed-loop system zeros).

A slightly simpler version of the criterion may be stated if, instead of using the function $D(s) = 1 + G(s)K(s)$, we draw the Nyquist plot of the open-loop transfer function $L(s) = G(s)K(s)$ and then count encirclements of the point $(-1, j0)$, rather than the origin. This gives the following modified form of the Nyquist criterion:

The number of unstable closed-loop poles (Z) is equal to the number of unstable open-loop poles (P) plus the net number of clockwise encirclements (N) of the point $(-1, j0)$ by the Nyquist plot of $L(s) = G(s)K(s)$, i.e.

$$Z = P + N$$

From the above, it is clear that a stable closed-loop system can only become unstable if the number of encirclements of the point $(-1, j0)$ by the Nyquist

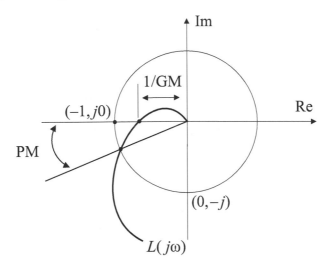

FIGURE 4.6: Nyquist plot with Gain Margin and Phase Margins.

plot changes. From this observation, we can define two important measures of *robust stability*, i.e. measures of the amount of uncertainty required to destabilise a stable closed-loop system, based on the the closeness with which the Nyquist plot passes to the point $(-1, j0)$. For example, in the case of an open-loop stable system, closed-loop stability requires that the Nyquist plot of $L(s)$ does not encircle the point $(-1, j0)$, i.e. that it crosses the negative real axis at a point between the origin and -1. As shown in Fig. 4.6, we can thus define the Gain Margin (GM) and Phase Margin (PM) for the system as:

$$\text{Gain Margin GM} = 20 \log\frac{1}{|L(j\omega_{cp})|} [dB]$$

$$\text{Phase Margin PM} = 180° + \arg\{L(j\omega_{cg})\}$$

where ω_{cp} is the phase crossover frequency, i.e. the frequency at which the phase of $L(j\omega) = 180°$, and ω_{cg} is the gain crossover frequency, i.e. the frequency at which the gain of $L(j\omega) = 1$. These margins provide measures of the amount of uncertainty in system gain and phase which can be tolerated by the closed-loop system before it loses stability. The Phase Margin also provides a measure of robustness to time delays in the feedback loop. This can be appreciated by noting that a time delay term $e^{-\tau_d s}$ has unity gain and phase equal to $-\omega\tau_d$ rads/s. Thus, for example, a closed-loop system with a PM of 35° (or $\frac{35\pi}{180} = 0.6109$ rads) can tolerate an additional time delay of $\tau_d = 0.6109/\omega_{cg}$ before losing stability. This also explains the lack of robustness of the system shown in Fig. 4.3 to even very small amounts of time delay. The phase margin of this system is only 10.2° or 0.1780 rads, with $\omega_{cg} = 44.5$

rads. Thus the maximum amount of time delay which may be tolerated in the closed-loop system is $0.178/44.5 = 0.004$, and a time delay of 5 ms causes instability, as shown.

Adequate Gain and Phase Margins represent minimal requirements for robust stability in feedback systems — typical values required in traditional control engineering applications are 6 dB of Gain Margin and 35° of Phase Margin. Their limitations as robustness measures are by now also widely recognised; however, among the most important of which are:

- Gain and Phase Margins do not measure robustness to *simultaneous* changes in system gain and phase, i.e. when calculating GM, the phase of the system is assumed to be perfectly known, and vice versa.

- In systems with multiple feedback loops Gain and Phase margins can be calculated for each loop one at a time, but may give unreliable results, since they do not take into account cross-coupling effects between different feedback paths.

- Finally, Gain and Phase Margins are defined for LTI systems, and do not take into account the potential destabilising effects of nonlinear or time-varying dynamics in feedback systems.

The above considerations have motivated the development of many more sophisticated robustness measures in recent control engineering research, and in Chapter 6 we provide more details of several of these tools and their application in the context of systems biology.

4.3   Performance of negative feedback systems

After discussing the stability of negative feedback systems, we now focus on the analysis of their performance, i.e. all those properties that determine the effectiveness of the closed-loop system response. Performance indices typically used in control engineering include

a) steady-state error of the output with respect to the reference signal;

b) response speed, measured in terms of rise time, settling time and bandwidth of the frequency response (see Section 2.8);

c) capability to reject disturbances;

d) amplitude and rate of variation of the control input signal required.

These characteristics can be analysed by studying the transfer functions between the exogenous inputs (reference signals $r(t)$ and disturbances $d(t)$)

and the variables of the system influenced by these inputs (control inputs $u(t)$, error signals $e(t)$ and controlled outputs $y(t)$).

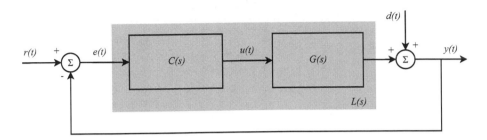

FIGURE 4.7: Block diagram of the classical negative feedback control loop.

Thus, with reference to the control scheme shown in Fig. 4.7, it is useful to define the following transfer functions:

- *Sensitivity Function*

$$S(s) = \frac{1}{1 + G(s)K(s)} \tag{4.5}$$

- *Complementary Sensitivity Function*

$$T(s) = \frac{G(s)K(s)}{1 + G(s)K(s)} \tag{4.6}$$

- *Control Sensitivity Function*

$$Q(s) = \frac{K(s)}{1 + G(s)K(s)} \tag{4.7}$$

The following relations link the Laplace transform of the input and output variables

$$Y(s) = T(s)R(s) + S(s)D(s) \tag{4.8a}$$
$$U(s) = Q(s)R(s) - Q(s)D(s) \tag{4.8b}$$
$$E(s) = S(s)R(s) - S(s)D(s) \tag{4.8c}$$

Note carefully that, without feedback control, there is no way that the effect of disturbance signals $d(t)$ on the output of the system $y(t)$ could be attenuated, *no matter what changes were made to the system $G(s)$.* Indeed, to obtain a perfect tracking of the reference signal, $r(t)$, along with a perfect rejection of the disturbance, $d(t)$, the conditions

$$T(j\omega) = 1 \tag{4.9}$$
$$S(j\omega) = 0 \tag{4.10}$$

should be ideally satisfied for all values of ω. Under these conditions, however, the relation $Q(s) = G(s)^{-1}T(s)$ yields $Q(s) = G(s)^{-1}$. Since in practice $G(s)$ is always such that when $s \to \infty$, $G(s) \to 0$, we have that $Q(s) \to \infty$. Hence, the control effort, i.e. the size of the control input signal, required to provide perfect tracking and disturbance rejection increases with frequency and eventually becomes unbounded. In physical systems, this relation places serious limits on the performance of feedback systems, since there are always practical limitations on the size (and rate of change) of control input signals (e.g. limits on the angular position and velocity of an aircraft rudder place limitations on the frequency of pilot reference inputs which may be tracked, and wind gust disturbances which may be attenuated, by the flight control system). In biological systems, changes in the concentrations of certain molecules, or in drug doses, will also be intrinsically limited by the availability of molecular compounds, diffusion effects, toxicological effects, etc. and thus there will always be limitations on the control performance which may be obtained.

It is also important to notice that the condition

$$S(s) + T(s) = 1 \qquad (4.11)$$

holds, and thus the frequency responses of $S(j\omega)$ and $T(j\omega)$ cannot be assigned independently. This reveals a fundamental tradeoff in the performance of a feedback system: since the function $T(s) \to 0$ when $s \to 0$, then $S(s) \to 1$ and the conditions in Eq. (4.9)–(4.10) are not realisable for $\omega \in [0, +\infty)$. Moreover, decreasing $T(j\omega)$ in a given interval of ω (to limit the size of the control input) causes an increase in the sensitivity of the system to disturbances, $S(j\omega)$, and vice versa.

For biological feedback systems, in contrast to engineered control systems, it is sometimes difficult to make a clear distinction between the *controller C(s)* and the *system being controlled G(s)* (although this distinction is more clear in the context of Synthetic Biology, where one might be interested in the design of the controller). To avoid this complication, in the following we will focus on the *open-loop transfer function L(s) = G(s)C(s)*. At this point we also make a clear distinction between the terms "regulation" and "tracking." Although the term regulation is often loosely used in biology to indicate any type of feedback control, it has a very precise meaning in the control engineering literature, i.e. the capability of a control system to keep a controlled variable at, or close to, the value of a *constant* reference input. Tracking, on the other hand, refers to the capability of a control system to follow dynamic changes in the reference input. When $r(t)$ is constant ($r(t) = r$), the controlled variable $y(t)$ in a negative feedback system should reach a value equal, or at least close to, r after a transient time interval. To see what conditions must be satisfied for this to occur, let us make explicit the gain and the number of poles at the origin of the loop transfer function, by expressing it as

$$L(s) = \frac{C}{s^n} L'(s),$$

with $L'(0) = 1$. Now consider a step reference input at time $t = 0$ with amplitude $\bar{r}$. The steady-state error can be computed by applying the initial value theorem*, which yields

$$e_{ss} = \bar{r} - \lim_{t \to \infty} y(t)$$

$$= \left(1 - \lim_{s \to 0} T(s)\right)\bar{r}$$

$$= \left(1 - \lim_{s \to 0} \frac{C/s^n}{1 + C/s^n}\right)\bar{r} = \lim_{s \to 0}\left(\frac{s^n}{s^n + C}\right)\bar{r}.$$

Hence, if $L(s)$ has no pole at the origin (i.e. $L(s)$ contains no integrators), the steady-state error is

$$\frac{1}{1 + C}\bar{r}.$$

If, however, $n \geq 1$, then $e_{ss} = 0$ regardless of the values of C or $\bar{r}$ — an extremely robust level of performance! This is the basis for the use of integral control in many industrial feedback systems. By using the controller to introduce an integrator into the feedback loop, the control system acts in such a way that the control effort is proportional to the integral of the error, and thus perfect steady-state tracking of step changes in the reference signal can be guaranteed. This fact can be generalised to different types of reference signals and takes the name *internal model principle*: in order for the closed-loop system to perfectly (that is with $e_{ss} = 0$) track an assigned reference signal, the loop transfer function must include the Laplace transform of such a signal. For example, if the reference signal is a ramp, $r(t) = t \cdot 1(t)$, it can be readily shown that $L(s)$ must contain at least two integrators in order to achieve perfect tracking.

So far we have not considered the disturbance $d(t)$, which of course affects the regulation error as well. This effect is described by the sensitivity function, according to Eq. (4.8a). Analogously to the regulation problem, perfect rejection of a step disturbance, $d(t) = \bar{d}1(t)$, requires $L(s)$ to exhibit at least a pole at the origin; indeed, the contribution of the disturbance to the steady-state error is

$$e_\infty = \lim_{t \to \infty} y(t)$$

$$= \lim_{s \to 0} S(s)\bar{d}$$

$$= \lim_{s \to 0} \frac{1}{1 + C/s^n}\bar{d} = \lim_{s \to 0}\frac{s^n}{s^n + C}\bar{d}.$$

The arguments above can be extended to other types of signals: to completely reject a disturbance whose Laplace transform is $\bar{d}/s^n$, $L(s)$ must include at

---

*The initial value theorem states that if $F(s) = \int_0^\infty f(t)e^{-st}dt$ then $Lim_{t \to 0}f(t) = Lim_{s \to \infty}sF(s)$.

least n poles in the origin; if the number of poles is $n-1$, the final tracking error will be equal to $\bar{d}/C$.

In Chapter 2, we have seen that the dynamic behaviour of a linear system, namely the rise time, settling time, overshoot and oscillatory nature of the response, are mostly determined by the poles of the transfer function. For the closed-loop system of Fig. 4.7, taking $L(s) = N_L(s)/D_L(s)$, we obtain

$$T(s) = \frac{N_L(s)}{D_L(s) + N_L(s)}, \tag{4.12}$$

and hence the poles are the roots of the polynomial $D_L(s) + N_L(s)$. These roots can be computed numerically or studied through the *root locus* method, which we will not discuss here. However, the most simple and effective way to gain some insight into closed-loop dynamic behaviour is to look at the frequency response of the loop transfer function. Assume that the frequency response $L(j\omega)$ has no unstable pole and $|L(j\omega)| = 1$ only at ω_c, which will be denoted as the *critical frequency*. If $|L(j\omega)|$ is high at low frequencies and rapidly decreases after the critical frequency, as in the example depicted in Fig. 4.8, we can state the following approximations

$$|1 + L(j\omega)| \approx |L(j\omega)|, \quad \omega < \omega_c$$
$$|1 + L(j\omega)| \approx 1, \quad \omega > \omega_c,$$

hence

$$|T(j\omega)| \approx \begin{cases} 1, & \omega < \omega_c \\ |L(j\omega)|, & \omega > \omega_c \end{cases} \tag{4.13}$$

The Bode diagrams of the magnitude of $L(j\omega)$ and $T(j\omega)$, shown in Fig. 4.8 for a typical case, confirm the validity of the approximations. Under these assumptions, the harmonic components of the reference signal at frequencies lower than ω_c are transferred to the output almost unchanged, whereas those beyond the critical frequency are attenuated. Therefore, the critical frequency represents a good approximation of the bandwidth of the closed-loop system, which is expected to exhibit a pair of complex conjugate dominant poles around the critical frequency. Indeed, the response of the closed-loop system can be approximately described by the transfer function

$$T_a(s) = \frac{\omega_n^2}{s^2 + 2\zeta\omega_n s + \omega_n^2}, \tag{4.14}$$

where $\omega_n = \omega_c$. Moreover, it is possible to show that the phase margin φ_m, defined in Section 4.2, is linked to the damping coefficient ζ of the complex poles by the formula

$$\zeta = \sin(\frac{\varphi_m}{2}). \tag{4.15}$$

Transfer function (4.14) enables us to estimate the overshoot and number of oscillations of the closed-loop system output when the reference signal undergoes a step change. Note that Eq. (4.14) does not contain any zero; however,

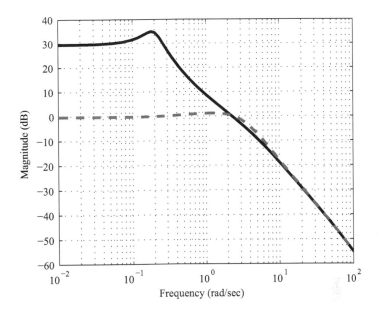

FIGURE 4.8: Frequency response of a loop transfer function $L(j\omega)$ (solid line) and its corresponding complementary sensitivity function $T(j\omega)$ (dashed line).

Eq. (4.12) shows that the zeros of $T(s)$ coincide with those of $L(s)$. It is important to take into account that low frequency zeros can significantly affect the step response, amplifying the initial overshoot and transient oscillations. We can also analyse the frequency response of the sensitivity function by making the same assumptions on $L(j\omega)$ as above, in order to derive the approximation

$$|S(j\omega)| \approx \begin{cases} \frac{1}{L(j\omega)}, & \omega < \omega_c \\ 1, & \omega > \omega_c \end{cases} \qquad (4.16)$$

which is confirmed by the example diagram shown in Fig. 4.9.

4.4 Fundamental tradeoffs with negative feedback

A recurrent theme in control systems research is the attempt to characterise fundamental limitations or tradeoffs between conflicting design objectives, since such information is invaluable to an engineer who is attempting to simultaneously satisfy many different stability and performance specifications. The identification of such properties in cellular networks could also provide

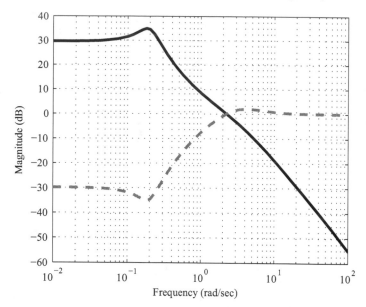

FIGURE 4.9: Frequency response of a loop transfer function $L(j\omega)$ (solid line) and its corresponding sensitivity function $S(j\omega)$ (dashed line).

deep insights into the design principles underlying the functioning of many different types of biological systems. In this section, we provide examples of some fundamental tradeoffs which hold exactly for linear negative feedback systems, and are likely to hold at least approximately for more general classes of systems.

One fundamental tradeoff which holds for negative feedback systems has already been given as Eq. (4.11) in the previous section. Consideration of the "shape" of the loop transfer function $L(s)$ provides further insight into the tradeoff between stability and performance in negative feedback systems. As discussed in the previous section, for accurate tracking of reference signals and good rejection of disturbances, $|L|$ should be large over the bandwidth of interest for the system. However, since the gain of most systems decreases at high frequency, large values of $|L|$ at high frequency require very large controller gains, and hence large, high-frequency control signals. Such signals are very difficult and/or expensive to generate — in a physical system, such as an aircraft rudder, they would require the use of very powerful, high performance servomotors, while in a cell, the generation of large, high-frequency fluctuations in molecular concentrations would be likely to impose a heavy energy load on the organism. For this reason, the magnitude of L is usually required to "roll-off" to a low value, above a certain critical frequency, as shown in Fig. 4.8. So far, so good, since all of the above requirements can be captured by making $|L|$ very large at low frequencies and very small at high frequencies. Unfortunately, the resulting need to make $|L|$ roll-off steeply at frequencies near the crossover region (frequencies between where $|L| = 1$ and $\angle L = -180°$) is not compatible with ensuring closed-loop stability, [3]. This is because the amount of phase lag in L is directly related to its rate of roll-off. For example, consider a loop transfer function of the form $L = 1/s^n$. In this case, the value of $|L|$ drops by $20 \times n$ dB when ω increases by a factor of 10. However, the phase associated with L is given by $\angle L = -n \times 90°$. Thus, if we wish to preserve a phase margin of $45°$, then we need that $\angle L > -135$ and thus n should not exceed 1.5.

Another fundamental constraint on the performance of negative feedback systems, known as *The Area Formula* relates to the magnitude of the sensitivity function $S = 1/(1 + L)$ at different frequencies, [4, 5]. Under the mild assumption that the relative degree (degree of the denominator minus degree of the numerator) of $L(s)$ is at least 2, the area formula gives that

$$\int_0^\infty log|S(j\omega)|d\omega = \pi(log\ e)\left(\sum \text{Re } p_i\right)$$

where p_i are the unstable poles of L. Consider, for example, the system

$$G(s) = \frac{1}{(s + 1)(s + 2)}$$

with a negative feedback controller $K(s) = 10$. The open-loop transfer function L is stable and has relative degree 2. Thus, the right-hand side of the area formula is equal to zero, and so if the sensitivity (on a log scale) is plotted against frequency (on a linear scale), then the positive area under the graph is equal to the negative area, as shown in Fig. 4.10. Thus, the improvement in tracking and disturbance rejection at some frequencies obtained by making S small must be paid for at others, where the effect of the feedback controller is actually to decrease the performance of the system. Of course, in the case of open-loop unstable systems, the situation is even worse, since there is now more positive than negative area. As suggested in [4], an intuitive explanation for this is that some of the feedback is being "used-up" in the effort to shift unstable poles into the left-half plane, and thus there is less available for the reduction of sensitivity. Alert readers will by now probably have thought of a "get-out clause" for the area formula: since only a conservation of area is required, why not pay for large reductions in sensitivity at some frequencies by making an arbitrarily small increase in $|S|$ spread over an arbitrarily large frequency range? Unfortunately, if the bandwidth of L is limited (and in reality it always is), then this is not possible, [6]. For example, if the open-loop bandwidth must be less than some frequency ω_1 (where $\omega_1 > 1$), such that

$$|L(j\omega)| < \frac{1}{\omega^2}, \quad \forall\ \omega \geq \omega_1$$

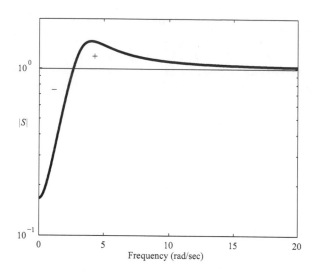

FIGURE 4.10: Illustration of the area formula

then for $\omega \geq \omega_1$

$$|S| \leq \frac{1}{1 - |L|} < \frac{1}{1 - \omega^{-2}} = \frac{\omega^2}{\omega^2 - 1}$$

and hence

$$\int_{\omega_1}^{\infty} log|S(j\omega)|d\omega \leq \int_{\omega_1}^{\infty} log\frac{\omega^2}{\omega^2 - 1}d\omega$$

The integral on the right-hand side of the above equation is finite, [6, 5], and so the available positive area at frequencies above ω_1 is limited. Thus, if $|S|$ becomes smaller and smaller over some part of the frequency range from zero to ω_1, then the required positive area must eventually be generated by making $|S|$ large at some other frequencies *below* ω_1.

In this section, we have provided only a few simple examples of the many different limitations which can be shown to apply to negative feedback systems in certain situations. Our analysis has been restricted to simple linear systems, and the reader might be entitled to question whether this type of analysis holds in general for biological systems. Two points need to be made here. The first is that results which hold for linear systems generally also hold for nonlinear systems when the deviations from the steady-state are small. Secondly, the type of analysis approach proposed here is extremely powerful because it provides hard bounds on system behaviour and so can be used to investigate the limits on performance of biological control systems. More generally, systems biology research is increasingly clarifying the crucial role of negative feedback in determining biological behaviour, and highlighting the similarities of such systems to engineered control systems. To take just one recent example, a study of the effects of negative feedback on three-tiered kinase modules in the MAPK/ERK pathway showed that the system recapitulates the design principles of a negative feedback amplifier, which is used in electronic circuits to confer robustness, output stabilisation and linearisation of nonlinear signal amplification, [7]. Directly analogous properties were observed in the biological behaviour of the MAPK/ERK as a result of negative feedback, which (i) converts intrinsic switch-like activation kinetics into graded linear responses, (ii) conveys robustness to changes in rates of reactions within the system and (iii) stabilises outputs in response to drug-induced perturbations of the amplifier.

4.5   Case Study V: Analysis of stability and oscillations in the p53-Mdm2 feedback system

Biology background: Tumour suppressor genes protect a cell from one step on the path to cancer. When such genes are mutated to cause a loss or reduction in their function, the cell can progress to cancer, usually in combination with other genetic changes. Whereas many abnormal cells usually undergo a type of programmed cell death (apoptosis), activated oncogenes can instead cause these cells to survive and proliferate. Most oncogenes require an additional step, such as mutations in another gene, or environmental factors, such as viral infection, to cause cancer. Cells which experience stresses such as DNA damage, hypoxia and abnormal oncogene signals activate an array of internal self-defense mechanisms. One of the most important of these is the activation of the tumour suppressor protein p53, which transcribes genes that induce cell cycle arrest, DNA repair and apoptosis. p53 transcriptionally activates the Mdm2 protein which, in turn, negatively regulates p53 by both inhibiting its activity as a transcription factor and by enhancing its degradation rate.

The negative feedback loop formed by p53 and Mdm2 also includes significant time delays arising from transcriptional and translational processes, and as a result can produce complex oscillatory dynamics. Oscillations of p53 and Mdm2 protein levels in response to ionising radiation (IR)-induced DNA damage appear to be damped in assays that measure averages over population of cells. Recent *in vivo* fluorescence measurements in individual cells, however, have shown undamped oscillations of p53 and Mdm2 lasting for at least 3 days. Although the oscillations are initially synchronised to the gamma irradiation signal, small variations in the timing of these oscillations inevitably arise due to stochastic variations across individual cells, causing the peaks to eventually go out of phase and thus the p53 and Mdm2 dynamics to appear as damped oscillations in assays over cell populations, [8].

Intriguingly, single-cell measurements in experiments with varying levels of IR have also revealed that increased DNA damage produces (on average) a greater number of oscillations, but has no effect on their average amplitude or period. The precise biological purpose of this "digital" type of response still remains to be fully elucidated, but one theory is that the oscillations of p53 may act as a timer for downstream events — genes inducing growth arrest (e.g. p21) are rapidly expressed during the first oscillation of p53, whereas proapoptotic p53 target genes such as Noxa, Puma or Bax are gradually integrated over multiple cycles of p53 pulses, ratcheting up at each pulse until they reach a certain threshold value that activates apoptosis, [9].

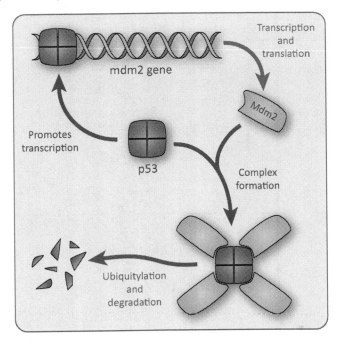

FIGURE 4.11: Biochemical interactions between p53 and Mdm2.

A block diagram of the p53-Mdm2 interactions is depicted in Fig. 4.11. Letting x_1 and x_2 represent the concentrations of p53 and Mdm2, the interaction dynamics can be approximated by the model

$$\dot{x}_1 = \beta_1\, x_1 - \alpha_{12}\, x_1\, x_2 \tag{4.17a}$$

$$\dot{x}_2 = \beta_2\, x_1(t-\tau) - \alpha_2\, x_2 \tag{4.17b}$$

which is based on the models presented in [8], with the parameter values given in Table 4.1. The dynamics of the intermediate biochemical reactions occurring after a change in the concentration of p53 are neglected so that only the final effect on the concentration of Mdm2 is considered. Therefore, the intermediate steps are represented in the model by means of a pure time delay τ. The presence of such a time delay can produce oscillations in the system response, as has been verified by experimental observations. On the other hand, the system does not oscillate for small values of τ. Thus, it is interesting to establish what is the minimum value of time delay for which the system exhibits undamped (or at least prolonged) oscillations.

An answer to this question can be found by applying tools from linear systems analysis, in particular the concept of phase margin, which was described in Section 4.2. Since system (4.17) is nonlinear, in order to apply this tool, we must derive a linearised model around an equilibrium point. By imposing

TABLE 4.1
Parameters values for system (4.17)

Parameter	Value	Unit	Description
β_1	2.3	h^{-1}	Self-induced generation rate coefficient for p53
β_2	24	h^{-1}	p53-induced generation rate coefficient for Mdm2
α_{12}	120	$x_{2\text{max}}^{-1} h^{-1}$	Mdm2 degradation rate coefficient
α_2	0.8	h^{-1}	Mdm2-induced degradation rate coefficient for p53

$\dot{x}_1 = 0$, $\dot{x}_2 = 0$ we get the equilibrium point

$$\bar{x}_1 = \frac{\alpha_2 \beta_1}{\alpha_{12} \beta_2}, \quad \bar{x}_2 = \frac{\beta_1}{\alpha_{12}}.$$

Thus, the linearised system is given by

$$\dot{\tilde{x}}_1 = (\beta_1 - \alpha_{12}\bar{x}_2)\, \tilde{x}_1 - \alpha_{12}\bar{x}_1\, \tilde{x}_2 \qquad (4.18a)$$

$$\dot{\tilde{x}}_2 = \beta_2\, \tilde{x}_1 - \alpha_2\, \tilde{x}_2 \qquad (4.18b)$$

where $\tilde{x}_i = x_i - \bar{x}_i$ for $i = 1, 2$. Now, we have seen in Section 4.2 that closed-loop stability can be inferred from the frequency response of the open-loop transfer function. The open-loop linearised system is obtained by deleting the feedback of $\tilde{x}_2$ and substituting it with an input signal $\tilde{u}$ in the first equation, which yields

$$\dot{\tilde{x}}_1 = (\beta_1 - \alpha_{12}\bar{x}_2)\, \tilde{x}_1 + \alpha_{12}\bar{x}_1\, \tilde{u} \qquad (4.19a)$$

$$\dot{\tilde{x}}_2 = \beta_2\, \tilde{x}_1 - \alpha_2\, \tilde{x}_2 \qquad (4.19b)$$

Note that the term containing the input $\tilde{u}$ is positive, because the minus sign is already included in the negative feedback scheme. The frequency response $L(j\omega)$ of system (4.19), given in Fig. 4.12, shows that the Phase Margin is equal to $32.7° = 0.57$ rad at $\omega_c = 1.24$ rad/s. Now recall that the maximum time delay the system can tolerate before losing stability can be computed as $\text{PM}/\omega_c = 0.57/1.24 = 0.46$ hours. This value can only be expected to be an approximate threshold, since it has been derived from a linear approximation of the nonlinear system: however, the smaller the perturbation from the equilibrium condition, the better the approximation will be. To test the validity of the computed delay threshold, we simulate the nonlinear system starting from the equilibrium condition and then inject a perturbation, by summing a square pulse signal $\tilde{d}(t) = \bar{d}\,(1(t) - 1(t - T))$ to x_2, for a number of different values of the time delay τ. Fig. 4.13 reports the time response of the nonlinear system for $\tilde{d} = \bar{x}_2/8$, $T = 10$ h (for better visualisation, the pulse perturbation is applied at time $t = 10$ h): it is clearly visible that the analysis conducted on the linearised system holds also for the nonlinear system, at least for a moderate perturbation of the state from the equilibrium condition. For $\tau < 0.46$ the oscillations induced by the perturbation dampen out, for $\tau = 0.46$ they exhibit a constant amplitude, whereas for $\tau > 0.46$ the oscillation is unstable. Note that in the latter case the oscillation amplitude does not grow unboundedly, but the system trajectory reaches a limit cycle (see Section 5.2).

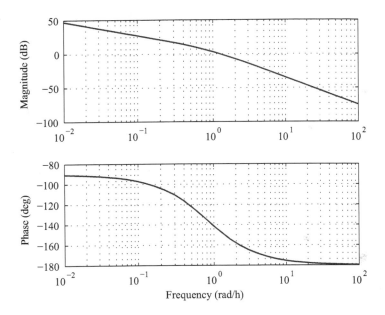

FIGURE 4.12: Frequency response of the loop transfer function $L(j\omega)$ of system (4.19).

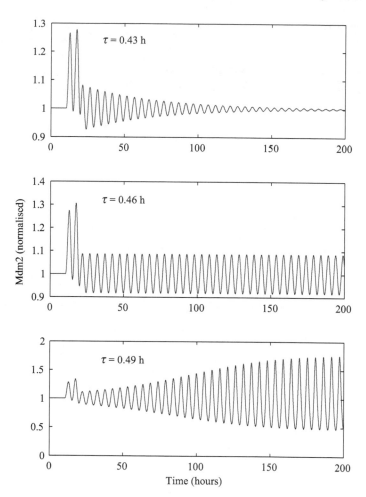

FIGURE 4.13: Simulations of the p53-Mdm2 system with different values of the time delay τ.

4.6   Case Study VI: Perfect adaptation via integral feedback control in bacterial chemotaxis

Biology background: Bacteria are constantly searching for sources of nutrients and trying to escape from locations containing harmful compounds. Bacteria like *E.coli* have an intricate locomotion system: each cell is endowed with several flagella, which can rotate clockwise (CW) or counter-clockwise (CCW). When rotating CCW the flagella are aligned into a single rotating bundle, therefore producing a movement along a straight line; CW rotation, on the other hand, causes unbundling of the flagella to create an erratic motion called a tumble. By alternating the rotational direction of the motor, *E. coli* swim through their environment in a sort of random walk. However, when a nutrient (e.g. aspartate) is sensed by the bacteria's membrane receptors, the random walk becomes biased towards the concentration gradient of the nutrients. This bias is achieved through controlling the length of time spent in CW and CCW rotation: when a bacterium recognises a change in the concentration of a nutrient, a signaling pathway is activated that eventually results in a prolonged period of CCW rotation. The same mechanism can be applied, by simply reversing the functioning logic, to flee from toxic compounds (e.g. phenol).

A key feature of this system is that the bacterium is very sensitive to changes in the concentration of the nutrient, but soon becomes insensitive to steady-state concentration levels. This is a sensible strategy, since if the surrounding environment contains a constant (either low or high) concentration of the nutrient, then there is no reason to swim in a particular direction. This property, which is very commonly encountered in biological sensing subsystems, is often referred to as *desensitisation* or *perfect adaptation*. It is the same mechanism, for example, that makes our olfactory system adapt to a constant odorant molecule concentration, eventually filtering it out.

The signaling pathway that underlies chemotaxis in *E. coli* has been thoroughly studied since the 1970's, [10, 11]. More recent studies have precisely characterised the bacterial perfect adaptation mechanism, using a mixture of computational modelling and experimental validation, [12, 13]. Furthermore, the results obtained using feedback control theory in [14] showed that the perfect adaptation encountered in bacterial chemotaxis stems from the presence of integral action in the signaling control scheme. This finding accounts for the high robustness of the chemotactic mechanism against large variations in molecular concentrations and environmental noise. Integral feedback has also been observed as a recurring motif in other biological systems that exhibit perfect adaptation, e.g. in calcium homeostasis, [15].

The sensing of chemical gradients by bacteria is mediated by transmembrane receptors, called methyl-accepting chemotaxis proteins (MCP). The binding of ligands to these MCP activates an intracellular signalling pathway, mediated by several *Che* proteins. The histidine kinase CheA is bound to the receptor via the adaptor protein CheW. CheA phosphorylates itself and then transfers phosphoryl groups to CheY. Phosphorylated CheY (CheY-P) diffuses in the cell, binds to the flagellar motors and induces CW rotation (cell tumbling). When an attractant binds to the MCP, the probability of the receptor being in the active state is decreased, along with the phosphorylation of CheA and CheY, eventually leading to a CCW flagellar rotation (straight motion). The probability of the chemoreceptors being active is also increased/decreased by adding/removing methyl groups, which is done by the antagonist regulator proteins CheR and CheB-P, respectively. CheB, in turn, is activated by CheA, by taking from the latter a phosphoryl group. The basic mechanism, which is captured by the computational model in [12], is that an increase in the ligand concentration is compensated for by increasing the methylation level. Since the two mechanisms have different time constants, the return to the original equilibrium requires a certain time interval. During this time interval, the system produces a transient response, corresponding to a reduction of the tumbling rate in favour of straight motion.

In the following, we present the mathematical model of bacterial chemotaxis developed in [12] and explain how it exhibits an integral feedback control structure, following the analysis in [14]. Additionally, we will show how the integral feedback property is crucially related to the biochemical assumption that the action of the methylation enzyme CheR is independent of the chemoattractant level.

4.6.1   A mathematical model of bacterial chemotaxis

The state variables and parameters included in the model are defined in Tables 4.2 and 4.3, [12]. When the number of methylation sites $M = 4$, the model comprises 26 state variables. Note that the concentration of the chemoattractant ligand, L, represents an exogenous input, whereas the concentrations of CheB$_P$ and CheR are assumed to be constant. The latter assumption is justified by the fact that methylation and demethylation are enzymatic reactions, in which the enzymes are not transformed.

The probability that a receptor is in its active state increases with the addition of methyl groups, whereas it is reduced by the binding of chemoattractant. The activation probability values are $\alpha_0^u = 0$, $\alpha_1^u = 0.1$, $\alpha_2^u = 0.5$, $\alpha_3^u = 0.75$, $\alpha_4^u = 1$ for unoccupied receptors and $\alpha_0^o = 0$, $\alpha_1^o = 0$, $\alpha_2^o = 0.1$, $\alpha_3^o = 0.5$, $\alpha_4^o = 1$ for occupied ones, where the subscript indicates the methylation level. Note that the unmethylated receptors are assumed to be always in the inactive state.

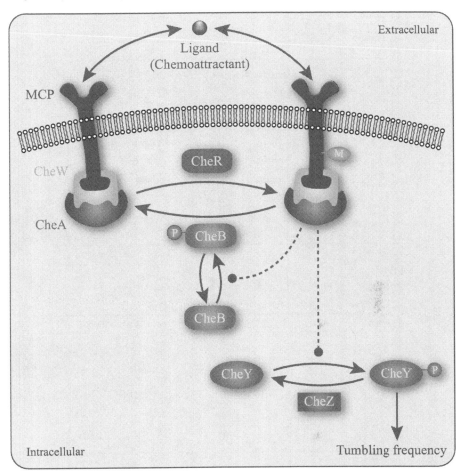

FIGURE 4.14: Chemotaxis regulation in response to variations in the concentration of chemoattractant.

The reactions considered in the model are

$$\bar{E}_m^\star \, (E_m^\star) + B \underset{d_b}{\overset{a_b(a_b')}{\rightleftharpoons}} \{E_m^\star B\} \overset{k_b}{\longrightarrow} E_{m-1}^\star, \quad m = 1, \ldots, M \qquad (4.20a)$$

$$\bar{E}_m^\star \, (E_m^\star) + R \underset{d_r}{\overset{a_r(a_r')}{\rightleftharpoons}} \{E_m^\star R\} \overset{k_r}{\longrightarrow} E_{m+1}^\star, \quad m = 0, \ldots, M-1 \qquad (4.20b)$$

$$E_m^u + L \underset{k_{-l}}{\overset{k_l}{\rightleftharpoons}} E_m^o, \qquad\qquad m = 0, \ldots, M. \qquad (4.20c)$$

The association kinetic constants of CheB$_P$, CheR with the receptor complex are a_b, a_r for the active form $\bar{E}_m^\star$ and a_b', a_r' for the inactive form $E_m^\star$, respec-

TABLE 4.2

State variables of the chemotaxis model (4.20)

State variable	Description
E_m^u	Receptor complex (MCP+CheW+CheA), with $m = 0, \ldots, M$ methyl groups, unoccupied by chemoattractant
E_m^o	Receptor complex (MCP+CheW+CheA), with $m = 0, \ldots, M$ methyl groups, occupied by chemoattractant
$\{E_m^\star B\}$	Receptor complex bound to CheB$_P$ ($\star$ can be u or o)
$\{E_m^\star R\}$	Receptor complex bound to CheR ($\star$ can be u or o)

TABLE 4.3

Parameters of the chemotaxis model (4.20) with perfect adaptation

Parameter	Value	Unit	Parameter	Value	Unit
a_b	800	1/(s μM)	d_r	100	1/s
d_b	1000	1/s	k_r	0.1	1/s
k_b	0.1	1/s	k_l	1000	1/(s μM)
a_r	80	1/(s μM)	k_{-l}	1000	1/s
a_r'	80	1/(s μM)	a_b'	0	1/(s μM)

tively. A key assumption in this model is that CheB can only associate with active receptors, denoted by $\bar{E}_m^\star$, and thus we assume $a_b' = 0$. Violation of this assumption affects the capability of the system to provide perfect adaptation, as will be demonstrated later. On the contrary, CheR can associate with both active and inactive receptors.

With respect to the schematic diagram of the overall system shown in Fig. 4.14, the mathematical model does not consider two mechanisms: a) the phosphorylation of CheY and its dephosphorylation by CheZ, and b) the spontaneous dephosphorylation of CheB and its phosphorylation by active CheA. These two subsystems are neglected in order to alleviate the computational burden and to focus the analysis on the regulatory mechanism that yields perfect adaptation. Mechanism a), indeed, acts as a transduction subsystem, by communicating the activation level to the flagellar motor, through the protein CheY$_P$. Note that it is not involved in any feedback loop and it can therefore be neglected in the analysis. Mechanism b), however, is implement-

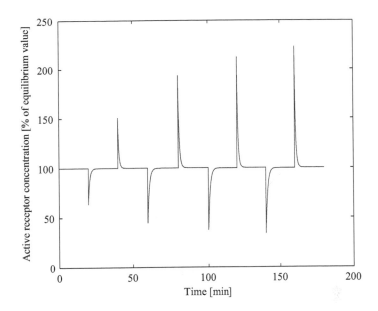

FIGURE 4.15: Concentration of active receptors in response to pulses of chemoattractant concentration: starting at $t = 20$ min, the ligand concentration is repeatedly set to a constant value for 20 minutes and then reset to zero for another 20 minutes, using different concentration levels (1,3,5,7 μM).

ing a feedback action: when the activation level increases, the concentration of CheB$_P$ increases as well, yielding a higher demethylation rate and, thus, counteracting the rise in the concentration of active receptors. Although this feedback action clearly plays an important role in the chemotaxis control system, it has been shown experimentally in [13] that it is *not* responsible for perfect adaptation, and therefore it is also neglected in our analysis.

By applying the law of mass action it is straightforward to translate Eq. (4.20) into a set of differential equations, e.g.

$$\frac{dE_1^o}{dt} = - a_b\, \alpha_1^o\, E_1^o\, CheB_P - a_b'\, (1 - \alpha_1^o)\, E_1^o\, CheB_P + d_b\, \{E_1^o B\} + k_b\, \{E_2^o B\}$$
$$- a_r\, \alpha_1^o\, E_1^o\, CheR - a_r'\, (1 - \alpha_1^o)\, E_1^o\, CheR + d_r\, \{E_1^o R\} + k_r\, \{E_0^o B\}$$
$$+ k_l\, E_1^u\, L - k_{-l} E_1^o\,, \tag{4.21}$$

where X denotes the concentration of species X. As demonstrated by the simulation results shown in Fig. 4.15, this model does indeed exhibit the perfect adaptation property encountered in wet lab experiments. However, the complexity of the model hampers the comprehension of the mechanisms underpinning such behaviour. To elucidate these mechanisms more clearly, we

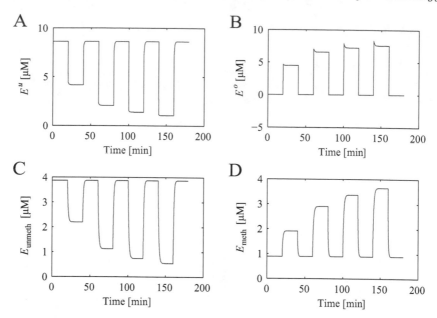

FIGURE 4.16: Changes in the concentrations of unoccupied (A), occupied (B), unmethylated (C) and methylated (D) receptors in response to 20 minute pulses of chemoattractant concentration.

must analyse in more detail the dynamics of the methylation/demethylation process.

4.6.2 Analysis of the perfect adaptation mechanism

The mechanism through which the system achieves perfect adaptation to changes in the ligand concentration cannot be seen explicitly from the reaction scheme in Fig. 4.14. At equilibrium, the rates at which receptors are being methylated and demethylated are equal. Recall that demethylation is assumed only to happen to activated receptors, and increased ligand binding decreases the probability of activation. This results in a very fast drop in the demethylation rate, due to the fast ligand binding dynamics. Because the methylation rate is constant, while the demethylation rate is reduced, the methylation level increases over time until the number of activated receptors returns to its original value and the system returns to equilibrium. At this point, the demethylation rate will also have returned to its original value, and the overall flux balance is restored. This mechanism is confirmed by the time courses reported in Fig. 4.16 which have been generated using the same pulses in chemoattractant concentration used in Fig. 4.15. Panels A and B show the very fast changes in the concentrations of unoccupied and occupied receptors which result in the fast deviations from equilibrium of the active receptor concentration shown in Fig. 4.15. Panels C and D show the (slower) changes in the concentrations of methylated and unmethylated receptors, which act to restore the active receptor concentration to its equilibrium value, as shown in Fig. 4.15.

To gain further insight into this intriguing feedback control system, let us explicitly write the balance equation for receptor methylation/demethylation, that is

$$\dot{z} = k_r \sum_{m=0}^{M-1} \{E_m^\star R\} - k_b \sum_{m=1}^{M} \{E_m^\star B\} , \qquad (4.22)$$

where $z := \sum_{m=1}^{M} E_m^\star$ is the total concentration of methylated receptors. This is not exactly the same as the total methylation level, which is actually given by the total concentration of bound methyl groups $\sum_{m=1}^{M} m \cdot E_m^\star$. However, in the following we use the scalar quantity z as an approximate indicator of the methylation level, in order to avoid the use of vector notation which would unnecessarily complicate the analysis. Since the methylation/demethylation reactions are assumed to follow Michaelis–Menten kinetics, we can substitute $\bar{E}_m^\star \cdot CheB_P/K_{Mb}$ for $\{E_m^\star B\}$, where K_{Mb} is the Michaelis–Menten constant of the demethylation reaction, given by $K_{Mb} = (k_b + d_b)/a_b$. Regarding the methylation rate, assuming that the protein CheR is present in small quantities with respect to the receptor, we can also assume that the concentration $\{E_m^\star R\}$ is almost equal to the total concentration of CheR, denoted by $CheR^T$. Thus, the methylation reaction constantly occurs at the maximum rate, equal to $k_r \, CheR^T$, and

$$\dot{z} = k_r \, CheR^T - \frac{k_b \, CheB_P}{K_{Mb}} \sum_{m=1}^{M} \bar{E}_m^\star$$

$$= k_r \, CheR^T - \Gamma \, \bar{E}, \qquad (4.23)$$

where $\Gamma = k_b \, CheB_P/K_{Mb}$ and $\bar{E}$ is the activity level (total concentration of active receptors) given by

$$\bar{E} := \sum_{m=1}^{M} \bar{E}_m^\star = \sum_{m=1}^{M} \alpha_m^u E_m^u + \sum_{m=1}^{M} \alpha_m^o E_m^o$$

Once again, to avoid having to use vector notation to represent the relative contribution of each methylated state, we approximate $\bar{E}$ with the function $f_E(L, z)$. This function depends on the ligand concentration L, because L determines the number of receptors which are unoccupied or occupied. It also depends on the methylation level z because z determines the relative numbers of receptors in each methylation state. Thus Eq. (4.23) can be represented as shown in Fig. 4.17 using the block diagram formalism, which effectively highlights the structural presence of an integral feedback control loop.

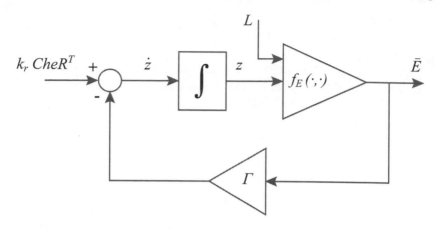

FIGURE 4.17:   Block diagram representation of the methylation-demethylation mechanism, showing the integral feedback control loop.

The system is at steady-state when the concentration of methylated receptors is constant, that is $\dot{z} = 0$; hence the active receptor concentration at steady-state, $\bar{E}_{ss}$, can be computed as

$$
\begin{aligned}
\bar{E}_{ss} &= \frac{k_r CheR^T}{\Gamma} \\
&= \frac{k_r\,CheR^T\,K_{Mb}}{k_b\,CheB_{Pss}} \\
&= \frac{k_r\,CheR^T\,K_{Mb}}{k_b\left(CheB^T - \sum_{m=1}^{M}\{E_m^\star B\}_{ss}\right)} \\
&= \frac{k_r\,CheR^T\,K_{Mb}}{k_b\,CheB^T - k_r\,CheR^T} = \frac{\gamma\,CheR^T\,K_{Mb}}{CheB^T - \gamma\,CheR^T},
\end{aligned}
$$

where $\gamma = k_r/k_b$ and we have exploited the fact that, by virtue of Eq. (4.22), at steady-state

$$
k_b \sum_{m=1}^{M}\{E_m^\star B\}_{ss} = k_r\,CheR^T.
$$

The expression for $\bar{E}_{ss}$ confirms that, as expected from our discussion of integral feedback control in Section 4.3, the concentration of active receptors at steady-state is independent of the ligand concentration, since it is uniquely determined by the total concentration of CheR and CheB, the constant K_{Mb} and the ratio of the kinetic constants k_r, k_b in the methylation and demethylation reactions.

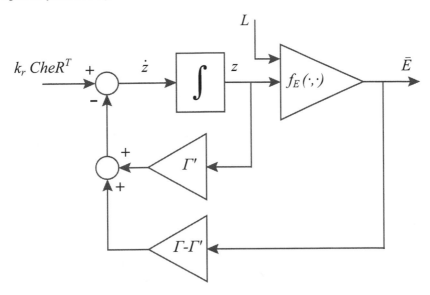

FIGURE 4.18: Block diagram representation of the methyla-tion/demethylation mechanism, assuming that CheB$_P$ can demethylate also nonactive receptors.

4.6.3 Perfect adaptation requires demethylation of active only receptors

In [12], it was recognised that the key assumption required in this model to obtain perfect adaptation is that CheB$_P$ can demethylate only active recep-tors. Although it has not been possible to directly confirm this assumption experimentally, some supporting evidence for it may be found in the litera-ture. For instance, in the face of a sudden increase of chemoattractant, the demethylation rate has been shown to fall sharply, [16],[17]. This could be explained by the sudden reduction in the number of active MCPs, caused by their association with the chemoattractant molecules.

Further confirmation of the necessity of this assumption can be provided by studying how the regulatory mechanism changes when the assumption is no longer valid. In this case, the kinetic constant a'_b for the association of CheB$_P$ with the inactive receptor $E^\star_m$ is no longer zero. Thus, Eq. (4.23) becomes

$$\dot{z} = k_r \, CheR^T - \frac{k_b \, CheB_P}{K_{Mb}} \bar{E} - \frac{k_b \, CheB_P}{K'_{Mb}} \left(z - \bar{E} \right)$$

$$= k_r \, CheR^T - (\Gamma - \Gamma') \, \bar{E} - \Gamma' \, z \,, \tag{4.24}$$

where $K'_{Mb} = (k_b + d_b)/a'_b$ and $\Gamma' = k_b \, CheB_P/K'_{Mb}$. Correspondingly, the control structure of Fig. 4.17 modifies to the one in Fig. 4.18. This block diagram shows that the control structure is now composed of two feedback loops, one on the methylation level and another on the activity level. Note that when $a_b' = a_b$ (i.e. CheB$_P$ can associate equally well with active and inactive receptors) then $\Gamma = \Gamma'$ and the activity level feedback loop vanishes. In this case the system would not be able to counteract the effect of changes in the ligand concentration on the activity level and only the methylation level would be regulated. If $0 < a_b' < a_b$, then the lower the value of a_b' the closer the system will be to the integral feedback structure of Fig. 4.18 and the more effective will be the adaptation mechanism.

The above arguments are confirmed by the simulations of the response of the system with different values of a_b', shown in Fig. 4.19.

The case study described above represents a striking example of how it is possible to support a biological hypothesis by rigorous engineering arguments: by exploiting the analysis tools of control theory, it has been possible to confirm that the chemotactic mechanism is based on the fact that CheB$_P$ demethylates only active receptors. In addition, we have been able to show that quasi-perfect adaptation can still be achieved when the demethylation of inactive receptors occurs at a very low rate.

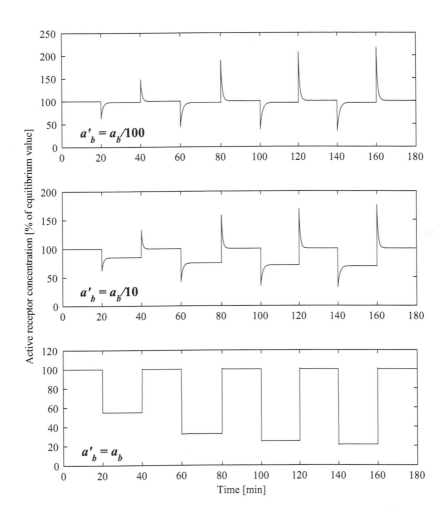

FIGURE 4.19: Concentration of active receptors in the chemotaxis model with non-perfect adaptation. The system is subjected to several pulses of chemoattractant concentration as in the experiment illustrated in Fig. 4.15. Quasi-perfect adaptation is achieved for low values of a'_b (top panel), whereas the system exhibits no adaptation when $a'_b = a_b$ (bottom panel).

References

[1] Franklin GF, Powell JD, and Emani-Naeini A. *Feedback Control of Dynamic Systems*. Boston: Addison-Wesley Publishing Company Inc., 3rd edition, 1994.

[2] Dorf RC. *Modern Control Systems*. Philadelphia: Prentice-Hall, 9th edition, 2000.

[3] Skogestad S and Postlethwaite I. *Multivariable Feedback Control*. Chichester: John Wiley, 2nd edition, 2005.

[4] Maciejowski JM. *Multivariable Feedback Design*. Boston: Addison-Wesley, 1989.

[5] Doyle JC, Francis BA, and Tannenbaum AR. *Feedback Control Theory*. New York: Macmillan, 1992.

[6] Freudenberg JS and Looze DP. Right half plane poles and zeros and design tradeoffs in feedback systems. *IEEE Transactions on Automatic Control*, AC-30:555–565, 1985.

[7] Sturm OE, Orton R, Grindlay J, Birtwistle M, Vyshemirsky V, Gilbert D, Calder M, Pitt A, Kholodenko B, and Kolch W. The mammalian MAPK/ERK pathway exhibits properties of a negative feedback amplifier. *Science Signalling*, 3(153):ra90, 2010.

[8] Geva-Zatorsky N, Rosenfeld N, Itzkovitz S, Milo R, Sigal A, Dekel E, Yarnitzky T, Liron Y, Polak P, Lahav G, and Alon U. Oscillations and variability in the p53 system. *Molecular Systems Biology*, doi:10.1038/msb4100068, 2006.

[9] Ma L, Wagner J, Rice JJ, Hu W, Levine AJ, and Stolovitzky GA. A plausible model for the digital response of p53 to DNA damage. *PNAS*, 102(4):14266–14271, 2005.

[10] Adler J and Tso W-W. Decision-making in bacteria: chemotactic response of Escherichia coli to conflicting stimuli. *Science*, 184(143):1292–1294, 1974.

[11] Macnab RM and Koshland DE. The gradient-sensing mechanism in bacterial chemotaxis. *PNAS*, 69:2509–2512, 1972.

[12] Barkai N and Leibler S. Robustness in simple biochemical networks. *Nature*, 387:913–917, 1997.

[13] Alon U, Surette MG, Barkai N, and Leibler S. Robustness in bacterial chemotaxis. *Nature*, 397:168–171, 1999.

[14] Yi T-M, Huang Y, Simon MI, and Doyle J. Robust perfect adaptation in bacterial chemotaxis through integral feedback control. *PNAS*, 97(9):4649–4653, 2000.

[15] El-Samad H, Goff JP, and Khammash M. Calcium homeostasis and parturient-hypocalcemia: an integral feedback perspective. *Journal of Theoretical Biology*, 214:17–29, 2002.

[16] Toews ML, Goy MF, Springer MS, and Adler J. Attractants and repellents control demethylation of methylated chemotaxis proteins in Escherichia coli. *PNAS*, 76:5544–5548, 1979.

[17] Stewart RC, Russell CB, Roth AF, and Dahlquist FW. Interaction of CheB with chemotaxis signal transduction components in *Escherichia coli*: modulation of the methylesterase activity and effects on cell swimming behavior. In *Proceedings of the Cold Spring Harbor Symposium on Quantitative Biology*, 53:27–40, 1988.

5

Positive feedback systems

5.1 Introduction

As seen in the previous chapter, negative feedback control loops play an important role in enabling many different types of biological functionality, from homeostasis to chemotaxis. When evolutionary pressures cause negative feedback to be supplemented with or replaced by positive feedback, other dynamical behaviours can be produced which have been used by biological systems for a variety of purposes, including the generation of hysteretic switches and oscillations, and the suppression of noise. Indeed, it has recently been argued that intracellular regulatory networks contain far more positive "sign-consistent" feedback and feed-forward loops than negative loops, due to the presence of hubs that are enriched with either negative or positive links, as well as to the non-uniform connectivity distribution of such networks, [1]. In the case studies at the end of this chapter we consider some of the types of biological functionality which may be achieved by positive feedback. First, however, we provide an introduction to some of the tools which are available to analyse these types of complex feedback control systems.

5.2 Bifurcations, bistability and limit cycles

5.2.1 Bifurcations and bistability

In Chapter 3, we have seen that nonlinear systems can exhibit multiple equilibria, each one being (either simply or asymptotically) stable or unstable. As can clearly be seen in Fig. 3.10, for example, the position of the equilibrium points, along with their stability properties and regions of attraction, determine in large part the trajectories in the state-space, i.e. the behaviour of the system.

On the other hand, nonlinearity also implies that the number and location of the equilibrium points, as well as their stability properties, vary with the parameter values. Therefore, it comes as no surprise that the behaviour of a

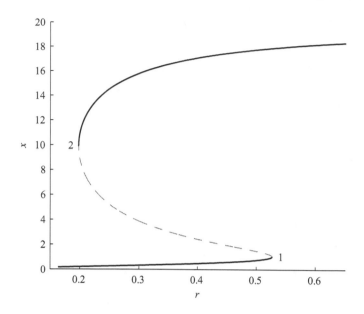

FIGURE 5.1: Bifurcation diagram of system (3.14).

nonlinear system might dramatically change when the value of some parameter varies, even by a small amount: this phenomenon is called a *bifurcation*.

In Example 3.2, we have shown that system (3.14) can have either one or three equilibrium points, depending on the values of the parameters r and q. Assume, for example, that the value of q is fixed at 20 and let r increase from 0.15 to 0.6. From Fig. 3.1 we see two bifurcation points, occurring at $r = 0.198$ and $r = 0.528$, where the number of equilibrium points changes from one (low value) to three and then back to one (high value).

A straightforward stability analysis, via linearisation at the equilibrium points, reveals that the low- and high-valued equilibrium points are always asymptotically stable, whereas the middle-valued one, when it exists, is unstable. The variations in the map of equilibrium points corresponding to changes of r can be effectively visualised by using a *bifurcation diagram*, in which the equilibrium values of some state variable are plotted against the bifurcation parameter. For example, the bifurcation diagram of system (3.14) is shown in Fig. 5.1: the solid lines represent the asymptotically stable equilibrium values, whereas the dashed line represents the unstable one. For intermediate values of r the system is *bistable*; it can evolve to the upper or lower branch of the diagram, depending on whether the initial condition is above or below the middle branch, respectively. The bifurcation diagram also informs us that there is a hysteresis-like behaviour in this system: when the system's state is on the lower stable equilibrium branch the state jumps to the higher stable equilibrium branch when r is increased beyond 0.528; however, to jump back to the lower stable condition, the value of r must drop below 0.198.

Bistability is a very important system-level property that is exhibited even by many relatively simple signalling networks. It is the mechanism that allows the production of switch-like biochemical responses, like those underlying commitment to a certain fate in the cell cycle and in the differentiation of stem cells, or the production of persistent biochemical "memories" of transient stimuli. Note that the presence of a hysteresis ensures a stable switching between the two operative conditions for the system; indeed, if the two thresholds were coincident, the system trajectories would constantly switch back and forth when the value of r is subject to stochastic variation around the bifurcation point.

Bifurcations can be classified according to the type of modifications they produce in the map of equilibrium points and in their stability properties. In the following we give a brief overview of the most common types of bifurcations, that is saddle node, transcritical and pitchfork, confining ourselves for simplicity to the case of first-order systems. For a comprehensive treatment of bifurcations and their applications to biological (and other) systems, the reader is referred to Strogatz's classical monograph [2].

Saddle-node bifurcation. This type of bifurcation occurs when there are two equilibrium points, one asymptotically stable and the other unstable. As the bifurcation parameter increases, the two points get closer and eventually collide, annihilating each other. The prototypical example of a saddle-node bifurcation is provided by the system

$$\dot{x} = r + x^2. \tag{5.1}$$

A dual bifurcation can be generated by changing the sign of the nonlinear term, that is

$$\dot{x} = r - x^2. \tag{5.2}$$

In the latter system, for small values of r there is a single stable equilibrium point, but as the parameter increases suddenly two equilibrium points appear (one asymptotically stable and the other unstable). For still higher values of r the system returns to having a single stable equilibrium point. The diagram in Fig. 5.1 thus exhibits two saddle-node bifurcations: as r increases, a pair of stable/unstable equilibrium points is generated at point 2 and destroyed at point 1.

Transcritical bifurcation. A transcritical bifurcation is characterised by an asymptotically stable and an unstable equilibrium point, which get closer together as the bifurcation parameter increases until they eventually collide and then separate, in the process exchanging their stability properties. The prototypical example of a transcritical bifurcation is provided by the system

$$\dot{x} = rx - x^2. \tag{5.3}$$

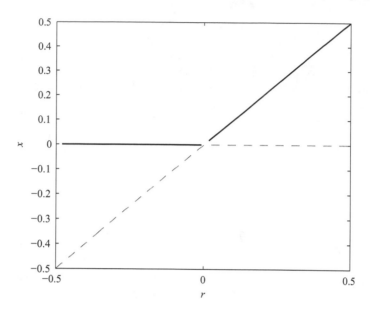

FIGURE 5.2: Transcritical bifurcation diagram of system (5.3).

which yields the bifurcation diagram shown in Fig. 5.2.

Pitchfork bifurcation.   A supercritical pitchfork bifurcation occurs when, as the bifurcation parameter increases, the asymptotically stable origin becomes unstable and, contemporarily, two new asymptotically stable equilibrium points are created, symmetrically with respect to the origin. This behaviour is exhibited, for example, by the system

$$\dot{x} = rx - x^3. \tag{5.4}$$

The associated bifurcation diagram is shown in Fig. 5.3(a), whereas Fig. 5.3(b) reports the dual case, termed a subcritical pitchfork bifurcation, which can be obtained from the system

$$\dot{x} = rx + x^3. \tag{5.5}$$

5.2.2   Limit cycles

A limit cycle is an isolated closed orbit which is periodically described by the state trajectory. The existence of periodic trajectories is not a prerogative of nonlinear systems; indeed, we have learned in Chapter 2 that oscillations arise, for example, when a linear system possesses a pair of purely imaginary eigenvalues. In the linear case, however, the amplitude of the oscillation

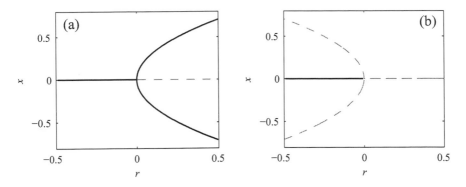

FIGURE 5.3: a) Supercritical pitchfork bifurcation diagram of system (5.4), b) subcritical pitchfork bifurcation diagram of system (5.5).

depends on the initial condition, which implies the presence of a family of periodic solutions: therefore, if the state is perturbed, the trajectory does not return to the original orbit. Moreover, the oscillations extinguish or diverge as soon as the real part of the eigenvalues slightly shift to the left or right half plane, respectively. This implies that the oscillations of linear systems are not robust to parameter uncertainties/variations, and therefore it is very unlikely that such systems can generate purely periodic trajectories in practice.

In nonlinear systems, on the other hand, limit cycles are independent of the initial conditions and neighbouring trajectories will be attracted to or diverge from a limit cycle (it will accordingly be termed a stable or unstable limit cycle, respectively). Thus, stable limit cycles are robust to state perturbations, i.e. they can exist in biological reality. In fact, the biological world is full of systems that produce periodic sustained oscillations, even for very long periods, for example, the mechanisms involved in the circadian clock, the cardiac pulse generator or the cell division cycle itself. Moreover, the uncertainties and disturbances which affect all biological processes suggest that the mechanisms generating such life-critical oscillations must be robust in the face of different initial conditions, parameter variations and environmental perturbations.

Focusing on the molecular level, it is worth mentioning the following result, taken from [3]: a necessary condition for exhibiting limit cycles, in a two species reaction system, is that it involves at least three reactions, among which one must be autocatalytic of the type

$$2X + \cdots \leftrightarrow 3X + \ldots$$

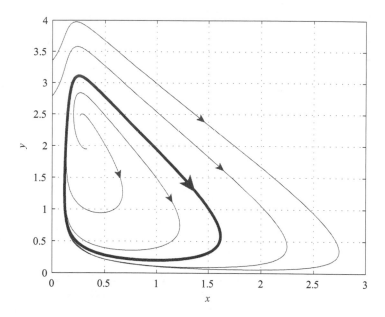

FIGURE 5.4: Phase plane of the simple chemical oscillator (5.7). The thick curve denotes the limit cycle.

Example 5.1

On the basis of the results above, among the possible candidates for chemical systems which exhibit limit cycles, the simplest such reaction mechanism can be shown to be, [4]:

$$X \underset{k_{1i}}{\overset{k_1}{\rightleftharpoons}} A, \quad B \overset{k_2}{\longrightarrow} Y, \quad 2X + Y \overset{k_3}{\longrightarrow} 3X. \tag{5.6}$$

Applying the law of mass action, the reaction kinetics are described by

$$\dot{x} = k_3 x^2 y + k_{1i} a - k_1 x \tag{5.7}$$

$$\dot{y} = k_2 b - k_3 x^2 y \tag{5.8}$$

The system exhibits a limit cycle for certain choices of the parameters, as shown by the phase plane in Fig. 5.4, which can be obtained with the parameters $k_1 = k_{1i} = k_2 = k_3 = 1$, $a = 0.1$, $b = 0.2$.   ⬚

Hopf bifurcation.   One further type of bifurcation which is relevant to the study of biological systems is the Hopf bifurcation. This bifurcation occurs

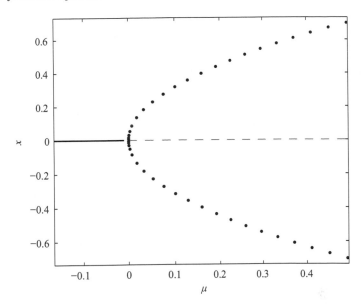

FIGURE 5.5: Supercritical Hopf bifurcation diagram of system (5.10).

when an asymptotically stable equilibrium mutates into an unstable spiral, i.e. a point in which the linearised system exhibits two unstable complex-conjugated eigenvalues, and the equilibrium is surrounded by a limit cycle. Therefore, when the bifurcation parameter surpasses the critical value, the system produces stable and robust oscillations. This is called a supercritical Hopf bifurcation, whereas the dual phenomenon, similarly to pitchfork bifurcations, is called a subcritical Hopf bifurcation. A prototypical second order system producing a Hopf bifurcation is

$$\dot{r} = \mu r - r^3 \tag{5.9a}$$
$$\dot{\theta} = \omega + br^2 \tag{5.9b}$$

where polar coordinates (r, θ) have been used. The same system can be translated in Cartesian coordinates, using the relations $x = r\cos\theta$, $y = r\sin\theta$, which yields

$$\dot{x} = \left[\mu - (x^2 + y^2)\right] x - \left[\omega + b(x^2 + y^2)\right] y \tag{5.10a}$$
$$\dot{y} = \left[\mu - (x^2 + y^2)\right] y + \left[\omega + b(x^2 + y^2)\right] x \tag{5.10b}$$

The supercritical Hopf bifurcation diagram of system (5.10) is depicted in Fig. 5.5, where the solid circles denote the amplitude of the oscillation.

5.3   Monotone systems

As discussed in Example 3.7 and shown in Fig. 3.10, biological systems which
exhibit more than one equilibrium can be analysed using standard graphical
approaches for the analysis of nonlinear systems in the phase plane. These
graphical methods are, however, generally only applicable to systems with
two states, which is clearly a significant limitation for the analysis of complex biological networks. In [5], a new method is described, based on the
theory of monotone systems, which allows the analysis of positive feedback
systems of arbitrary order for the presence of bistability or multistability (i.e.,
more than two alternative stable steady-states), bifurcations and associated
hysteretic behavior. The method relies on two conditions that are frequently
satisfied even in complicated, realistic models of cell signalling systems: monotonicity and the existence of steady-state characteristics. Below, we provide
an introduction to this approach, which will be used in Case Study VII to
analyse the dynamics of a positive feedback loop in a MAPK cascade.

The approach works by considering the positive feedback system in open-
loop, so that it can be described using the general set of ordinary differential
equations:

$$\dot{x}_1 = f_1(x_1, ..., x_n, \omega)$$
$$\dot{x}_2 = f_2(x_1, ..., x_n, \omega)$$
$$\vdots$$
$$\dot{x}_n = f_n(x_1, ..., x_n, \omega)$$

where $x_i(t)$ describes the concentration of some molecular species over time,
f_i is a differentiable function and ω represents an external input signal that
may be applied to the system. Assume that the output of the system is given
by some differentiable function of x, i.e. $\eta = h(x)$ (in practice, η will often
simply be one of the state variables, so that $\eta = x_i$). Thus η defines which
state variable, or combination of state variables, is fed back to the input of
the system via the positive feedback loop. In the following we assume for
simplicity that ω and η are both scalar, although extensions of the theory for
vector inputs and outputs have also been derived, [5].

In order to apply the test for multistability developed in [5], the system
defined above must satisfy two critical properties: (A) the open-loop system
has a monostable steady-state response to constant inputs, i.e. the system
has a well-defined steady-state input/output (I/O) characteristic; and (B) the
system is strongly I/O monotone, i.e. there are no possible negative feedback
loops, even when the system is closed under positive feedback.

Property A means that, for any constant input signal $\omega(t) = a$ for $t > 0$
(i.e. a step-function input stimulus), and for any initial conditions $x_1(0), y_1(0)$,
the solution of the above system of differential equations converges to a unique

steady-state, which depends on the particular step magnitude a, but not on the initial states. When this property holds, $k_{x,y}(a)$ indicates the steady-state vector $\lim_{t \to +\infty}[x_1(t), y_1(t)]$ corresponding to the signal $\omega(t) = a$, and $k_\eta(a)$ indicates the corresponding asymptotic value $\eta(+\infty)$ for the output signal.

Property B (monotonicity) refers to the graphical structure of the interconnections between the dynamic variables in the system. This structure is described by the incidence graph of the system, which has $n + 2$ nodes, labeled ω, η and x_i, $i = 1, ..., n$. To create the incidence graph, a labeled edge (an arrow with a $+$ or $-$ sign attached to it) is drawn whenever a variable x_i (or input ω) directly affects the rate of change of a variable x_j, $j \neq i$ (or the value of the output η). A $+$ sign is attached to each label whenever the effect is positive and a $-$ sign when the effect is negative. By definition, no edges are drawn from any x_i to itself. Thus, if $f_i(x, \omega)$ is strictly increasing with respect to x_j for all (x, ω), then a positive edge is drawn directed from vertex x_j to x_i, while if $f_i(x, \omega)$ is strictly decreasing as a function of x_j for all (x, ω), then a negative edge is drawn directed from vertex x_j to x_i. If f_i is independent of x_j, no edge from x_j to x_i is drawn. The same procedure is followed for edges from the vertex ω to any vertex x_j, and from any x_j to η. If an effect is ambiguous, because it depends on the actual values of the input or state variables, such as in the example $\dot{x}_1 = (1 - x_1)x_2 + \omega$, where $f_1(x_1, x_2, \omega) = (1 - x_1)x_2 + \omega$ is an increasing function of x_2 if $x_1 < 1$, but is a decreasing function of x_2 if $x_1 > 1$, then a graph cannot be drawn and the method as described here does not apply. The sign of a path (the individual edges transversed in any direction, forwards or backwards) is then defined as the product of the signs along it, so that the corresponding path is simply called positive or negative. A system is said to be strongly I/O monotone (i.e. it satisfies property B) provided that the following four conditions hold for the incidence graph of the system:

1. Every loop in the graph, directed or not, is positive.

2. All of the paths from the input to the output node are positive.

3. There is a directed path from the input node to each node x_i.

4. There is a directed path from each x_i to the output node.

Note that conditions (1) and (2) together amount to the requirement that every possible feedback loop in the system is positive — properties (3) and (4) are technical conditions needed for mathematical reasons.

If the system can be shown to satisfy both properties A and B, then it can be analysed for the property of bistability as follows. Graph together the characteristic k_η, which represents the steady-state output η as a function of the constant input ω, with the diagonal $\eta = \omega$. Algebraically, this amounts to looking for fixed points of the mapping k_η. If the characteristic k_η is sigmoidal, as shown in Fig. 5.6, then there will be three intersections between these graphs, which we label points I, II and III, respectively. Note that the

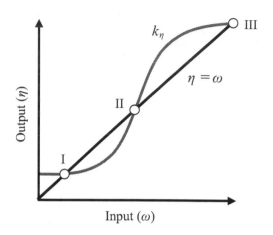

FIGURE 5.6: The sigmoidal steady-state I/O static characteristic curve k_η has three intersections with the line representing ω as a function of η for unitary positive feedback. The three intersection points (I, II, and III) represent two stable steady-states (I and III) and one unstable steady-state (II) for the closed-loop system.

slope of the characteristic k_η is < 1 at points I and III and > 1 at point II. If the open-loop system is now closed using unity positive feedback (i.e. by setting $\omega = \eta$), then it can be shown, [6], that the resulting closed-loop system has three equilibria, x_I, x_{II} and x_{III}, corresponding to the I/O pairs associated with the points I, II and III, respectively. The equilibria x_I and x_{III}, which correspond to the points at which the characteristic has slope < 1, are stable, whereas x_{II} is unstable, so that every trajectory in the state-space, except possibly for an exceptional set of zero measure, converges to either x_I or x_{III}, i.e. the system is bistable.

Note that if the characteristic k_η had not been sigmoidal, then there could not be three intersections, and the system could not be bistable *for any feedback strength*. Importantly, it is straightforward to show that any cascade composed of subsystems, each of which is monotone and admits a well-defined characteristic, will itself be monotone and admit a characteristic, [7]. Thus, in contrast to traditional phase-plane analysis, the approach described above can be applied to arbitrarily high-order systems. Finally, although the development above assumed the simple case where the output feeds back directly to the input, more complicated feedback loops may also be studied using the same basic approach, by a reduction to unity feedback, [5].

The computational analysis method described above also suggests an experimental approach to the detection of bistability in positive feedback systems.

If the feedback can be blocked in such a system, and if the feedback-blocked system is known (or correctly intuited based on biological insight) to be monotone, then if the experimentally determined steady-state stimulus-response curve of the feedback-blocked system is sigmoidal, the full feedback system is guaranteed to be bistable for some range of feedback strengths. Conversely, if the open-loop system exhibits a linear response, a Michaelian response, or any response that lacks an inflection point, the feedback system is guaranteed to be monostable despite its feedback. Thus, some degree of "cooperativity" or "ultrasensitivity" appears to be essential for bistability in monotone systems of any order.

5.4 Chemical reaction network theory

In this section, we introduce a powerful analysis tool named Chemical Reaction Network Theory (CRNT) [8],[9], which provides an alternative strategy, with respect to the approach presented in the previous section, to investigate the bistability of biomolecular systems. It is worth noting that the two approaches are complementary: CRNT is applicable to systems for which it is not possible to define a signed incidence graph. On the other hand, the Monotone Systems approach can cope with different types of kinetics, whereas the most useful results of CRNT are given for the special case of mass action kinetics.

The advantage of CRNT is that it provides a straightforward way to analyse the type of dynamical behaviour that one can expect from an arbitrarily complex network of chemical reactions, just by inspection of the topology of the associated graph. More specifically, CRNT enables us to establish whether an assigned reaction network can exhibit one or multiple equilibrium points, without even the need to write down the kinetic equations and assign values to the kinetic parameters. This point makes CRNT especially suitable for dealing with biomolecular systems, whose parameters are often unknown or subject to significant variability among different individuals.

Although CRNT is not a standard topic in the field of control engineering, it is closely related to it, since it deals with the study of equilibrium points and their stability properties. Moreover, it is becoming increasingly popular as a tool for systems biologists, for example as a method to sift kinetic mechanism hypotheses [10] and to study multistability in gene regulatory networks [11]. Thus, in view of the relationship discussed in the previous sections between positive feedback and bistability, it is appropriate to provide here at least an introductory overview of CRNT as an analysis tool for biological systems.

5.4.1   Preliminaries on reaction network structure

To facilitate the introduction of some preliminary definitions, we will refer to a simple example network, whose standard reaction diagram is as follows:

$$A_1{+}A_2 \rightleftharpoons A_3 \longrightarrow A_4{+}A_5 \rightleftharpoons A_6$$

$$2A_1 \longrightarrow A_2{+}A_7$$

$$A_8$$

We shall denote by N the number of species in the network under consideration, so for our example $N = 8$. With each species we associate a vector e_i, where $\{e_1, \ldots, e_N\}$ is the standard basis for $\mathbb{R}^N$, that is

$$e_1 = \begin{pmatrix} 1 \\ 0 \\ 0 \\ \vdots \\ 0 \end{pmatrix}, \quad e_2 = \begin{pmatrix} 0 \\ 1 \\ 0 \\ \vdots \\ 0 \end{pmatrix}, \quad \cdots, \quad e_N = \begin{pmatrix} 0 \\ 0 \\ 0 \\ \vdots \\ 1 \end{pmatrix}.$$

The *complexes* of a reaction network are the objects that appear before and after the reaction arrows. The number of distinct complexes will be denoted by n; thus in our network there are $n = 7$ complexes, namely $A_1 + A_2$, A_3, $A_4{+}A_5$, A_6, $2A_1$, $A_2{+}A_7$, A_8. With each reaction we shall associate a *reaction vector*, which is derived from the vectors e_i by summing the vectors associated with the products and subtracting those associated with the reactants, each multiplied by the respective stoichiometric coefficient. For example, for the reaction

$$A_1 + A_2 \to A_3 \tag{5.11}$$

the reaction vector is $r_1 = e_3 - e_1 - e_2 = \begin{pmatrix} -1 & -1 & 1 & 0 & 0 & 0 & 0 & 0 \end{pmatrix}^T$ and for

$$2A_1 \to A_2 + A_7 \tag{5.12}$$

we get $r_6 = e_2 + e_7 - 2e_1 = \begin{pmatrix} -2 & 1 & 0 & 0 & 0 & 0 & 1 & 0 \end{pmatrix}^T$. The reaction vectors span a linear subspace $\mathcal{S} \in \mathbb{R}^N$ which is called the *stoichiometric subspace*. The matrix $S = \begin{bmatrix} r_1 & r_2 & \cdots & r_p \end{bmatrix}$, where p is the number of reactions, is termed the *stoichiometric matrix* and is the starting point for various mathematical techniques used to determine network properties, especially in the study of metabolic networks, [12]. We shall say that a reaction network has *rank s* if the stoichiometric matrix has rank s. Recall that this amounts to stating that there exist at most $s \leq p$ linearly independent reaction vectors. The stoichiometric subspace enables us to characterise all the points of the state-space which are reachable by the system in terms of *stoichiometric compatibility classes*. We say that two points of the state-space, x' and x'', are stoichiometrically compatible if $x' - x''$ lies in S. At this point, we can partition the set of all positive state vectors into positive stoichiometric compatibility classes. In particular, the positive stoichiometric compatibility class containing $x \in \mathbb{P}^N$, where $\mathbb{P}^N$ is the positive orthant of $\mathbb{R}^N$, is the set $(x + S) \cap \mathbb{P}^N$, that is the set of vectors in $\mathbb{P}^N$ obtained by adding x to all vectors of S.

Looking at the standard reaction diagram (where each complex appears only once) of our example network, we readily notice that it is composed of two separate pieces, one containing the complexes $\{A_1 + A_2, A_3, A_4 + A_5, A_6\}$, the other containing the complexes $\{2A_1, A_2 + A_7, A_8\}$. There is no link between complexes of the two sets; therefore each set is called a *linkage class* of the network and the symbol l will be used to indicate the number of linkage classes in a network (in our case $l = 2$). Note that a linkage class is just a set of complexes, without any information about the related reactions. Two different complexes in a reaction network are *strongly linked* if there exist two directed arrow pathways, one pointing from one complex to the other and one in the reverse direction. By convention, every complex is considered strongly linked to itself. By a *strong linkage class* in a reaction network we mean a set of complexes such that each pair in the set is strongly linked to a complex that is not in the set. Note that the number of strong linkage classes does depend on the specific reaction diagram. A *terminal strong linkage class* is a strong linkage class containing no complex that reacts to a complex in a different strong linkage class. In rough terms, a strong linkage class is terminal if there is no exit from it along a directed arrow pathway. Each linkage class must contain at least one terminal strong linkage class; therefore, if we indicate by t the number of terminal strong linkage classes, then $t \geq l$.

An interesting result is that any two reaction networks with the same complexes and the same linkage classes also have the same rank. Hence, given only the complexes of a network and a specification of how they are partitioned into linkage classes, we can calculate the rank while ignoring the actual reaction topology. Indeed, to determine the rank, we can use any reaction network formed by the same complexes and linkage classes. A simpler network with n complexes and l linkage classes is one that contains only $p = n - l$ reactions; therefore its rank cannot exceed $n - l$ and the same holds for any network with the same number of complexes and linkage classes, no matter how many reactions it contains. Hence, we can state that the *deficiency* of a network, defined as

$$\delta = n - l - s \tag{5.13}$$

is always a nonnegative integer.

To understand CRNT, we also need the notion of a (weakly) reversible network: a *reversible* network is one in which each reaction is accompanied by its reverse. A network is *weakly reversible* if, whenever there exists a directed arrow pathway (consisting of one or more reaction arrows) pointing from one complex to another, there also exists a directed arrow pathway pointing from the second complex back to the first. The class of (weakly) reversible networks is a subset of the set of networks for which $t = l$.

5.4.2   Networks of deficiency zero

In this and in the next section we will provide two fundamental results in CRNT, which can be used to derive qualitative information about the trajectories of a system of nonlinear differential equations associated with a reaction network. The application of these results does not require a deep understanding of CRNT, but only some familiarity with the above preliminary notions concerning the complexes, rank, linkage classes and deficiency of a reaction network. In particular we will examine the case of networks of deficiency zero and of deficiency one.

It is important to remark that the results given below are general, for they apply to networks of any size and complexity, possibly involving hundreds of species and reactions. First let us consider the case of networks of deficiency zero. For any reaction network of deficiency zero the following statements hold true:

(i) If the network is not weakly reversible, then, for arbitrary kinetics (not necessarily mass action), the differential equations for the corresponding reaction system cannot admit a positive steady-state.

(ii) If the network is not weakly reversible, then, for arbitrary kinetics (not necessarily mass action), the differential equations for the corresponding reaction system cannot admit a cyclic state trajectory along which all species concentrations are positive.

(iii) If the network is weakly reversible, then, for mass action kinetics (but regardless of any particular positive value for the rate constants), the differential equations for the corresponding reaction system have the following properties: there exists within each positive stoichiometric compatibility class precisely one steady state; the steady state is asymptotically stable, and there is no nontrivial cyclic state trajectory along which all species concentrations are positive.

Precluding that the network can admit a positive steady-state means that if some steady-state exists it must be such that at least certain species concentrations are zero. Note also that the above result does not entirely preclude the existence of nontrivial cyclic state trajectories. For arbitrary kinetics there might be cyclic state trajectories such that some concentrations are always zero. When mass action kinetics are assumed, instead, it is possible to show that the system cannot generate any nontrivial cyclic composition trajectory.

Example 5.2

Let us illustrate the applicability of the above result by means of an example. Consider again our example reaction network — we want to establish whether this system admits a positive steady-state or a cyclic trajectory along which all species concentrations are positive. The network exhibits zero deficiency and is not weakly reversible, as is readily seen by considering the reaction path from complex $A_1 + A_2$ to A_6, for which there exists no reverse pathway. Therefore, according to statements (i) and (ii) above, we can rule out the existence of a positive steady-state or a cyclic trajectory (along which all species concentrations are positive) regardless of either the kinetics assigned to the reactions or the values of the parameters.

It is interesting to see what happens if we slightly modify our simple network, by making the reaction $A_3 \rightarrow A_4 + A_5$ reversible, as shown in the following reaction diagram:

$$A_1{+}A_2 \underset{k_2}{\overset{k_1}{\rightleftharpoons}} A_3 \underset{k_4}{\overset{k_3}{\rightleftharpoons}} A_4{+}A_5 \underset{k_6}{\overset{k_5}{\rightleftharpoons}} A_6$$

$$2A_1 \overset{k_7}{\longrightarrow} A_2{+}A_7$$

$$\underset{k_8}{\searrow} \quad \underset{A_8}{\overset{k_9}{\nearrow}} \overset{k_{10}}{\nearrow}$$

This modification renders the network weakly reversible. Note that the complexes and the linkage classes of the modified network are the same as the original one, and therefore the deficiency of the modified network equals zero. If we assume that the reaction kinetics are all of mass action type, the system is described by the following system of differential equations

$$\dot{c}_1 = -k_1 c_1 c_2 + k_2 c_3 - 2k_7 c_1^2 + k_8 c_8 \tag{5.14a}$$

$$\dot{c}_2 = -k_1 c_1 c_2 + k_2 c_3 + k_7 c_1^2 + k_9 c_8 - k_{10} c_2 c_7 \tag{5.14b}$$

$$\dot{c}_3 = k_1 c_1 c_2 + k_4 c_4 c_5 - (k_2 + k_3) c_3 \tag{5.14c}$$

$$\dot{c}_4 = k_3 c_3 - (k_4 + k_5) c_4 c_5 + k_6 c_6 \tag{5.14d}$$

$$\dot{c}_5 = k_3 c_3 - (k_4 + k_5) c_4 c_5 + k_6 c_6 \tag{5.14e}$$

$$\dot{c}_6 = k_5 c_4 c_5 - k_6 c_6 \tag{5.14f}$$

$$\dot{c}_7 = k_7 c_1^2 + k_9 c_8 - k_{10} c_2 c_7 \tag{5.14g}$$

$$\dot{c}_8 = -(k_8 + k_9) c_8 + k_{10} c_2 c_7 \tag{5.14h}$$

where the i-th state variable, c_i, is the concentration of species A_i. To study the behaviour of system (5.14) we can apply statement (iii) above, which allows us to conclude that, regardless of the (positive) value of the kinetic parameters, the system admits precisely one positive steady-state, which is asymptotically stable. Moreover the system does not admit a periodic state trajectory along which all species concentrations are positive. It is not difficult to see that providing such definitive answers to these questions by using other mathematical approaches would have been extremely difficult. ☐

The results discussed above are for networks of deficiency zero; however, there are a number of other interesting propositions and remarks that extend these basic results and provide more specific information. For example, CRNT allows us to state that, given a deficiency zero network containing the null complex, the corresponding system (no matter the reaction kinetics) admits no steady-state at all if the null complex does not lie in a terminal strong linkage class. The interested reader is referred to [8] for additional results.

5.4.3   Networks of deficiency one

Let us now consider the case of networks of deficiency one. In contrast to the results in the previous section, the results provided by CRNT for this case give no dynamical information: they are only concerned with the uniqueness and existence of positive steady-states. Also, for networks of nonzero deficiency, the lack of weak reversibility no longer precludes the existence of (multiple) positive steady-states. Indeed, the weak reversibility condition is replaced by the far milder condition that each linkage class contain no more than one terminal strong linkage class.

To better understand the following result it is important to note that the deficiency of a reaction network need not be the same as (in fact it is always greater or equal than) the sum of the deficiencies of its linkage classes. It is also important to point out that the following result holds for networks where the deficiencies of the individual linkage classes are less than one, but this does not mean that the deficiency of the entire network must be less than one.

Consider a mass action system for which the underlying reaction network has l linkage classes, each containing just one terminal strong linkage class. Suppose that the deficiency of the network and the deficiencies of the individual linkage classes satisfy the following conditions:

(i) $\delta_\theta \leq 1$, $\theta = 1, 2, \ldots, l$

(ii) $\sum_{\theta=1}^{l} \delta_\theta = \delta$.

Then, no matter what (positive) values the rate constants take, the corresponding differential equations can admit no more than one steady-state within a positive stoichiometric compatibility class. If the network is weakly reversible, the differential equations admit precisely one steady-state in each positive stoichiometric compatibility class.

For networks having just one linkage class, condition (ii) above is satisfied trivially. Thus, the following result is also readily derived: A mass action system for which the underlying reaction network has just one linkage class can admit multiple steady-states within a positive stoichiometric compatibility class only if the deficiency of the network or the number of its terminal strong linkage classes exceeds one.

The above result represents a generalisation of the previous result for networks of deficiency zero; indeed, it is concerned with the existence and uniqueness of one steady-state. Note also that it does not allow us to say anything about networks of deficiency one where all the linkage classes are of deficiency zero. This is a serious weakness, because deficiency one networks can exhibit multiple positive steady-states and we would like to have a tool to establish when this occurs. Fortunately, CRNT addresses this issue, at least for reaction networks with mass action kinetics, through the Deficiency One Algorithm, [9]. Given a deficiency one network satisfying certain weak *regularity* conditions, CRNT will indicate either that there does exist a set of rate constants such that the corresponding mass action differential equations admit multiple positive steady-states or else that no such rate constants exist. In the affirmative case, the algorithm will also provide a set of values of the kinetic parameters for which the system is multistable.

The detailed illustration of this aspect of CRNT goes beyond the scope of this book, as it would require the presentation of a number of new definitions and results and of a rather involved algorithm. Fortunately, it is not necessary to understand every detail of the theory to apply it: the algorithm is coded in the CRNT Toolbox*, which is freely available and easy to use. Using this toolbox, it is sufficient to fill in the network's reactions and run the algorithm to get a comprehensive report elucidating all the properties of the network that can be analysed by CRNT and, in particular, whether the corresponding dynamical system has multiple positive steady-states.

---

*http://www.chbmeng.ohio-state.edu/~feinberg/crnt

5.5 Case Study VII: Positive feedback leads to multistability, bifurcations and hysteresis in a MAPK cascade

Biology background: *Xenopus* oocytes are eukaryotic cells that undergo the classical steps of meiotic cell division. After the G1 and S phases, they carry out the early events of meiotic prophase: their homologous chromosomes pair up and undergo recombination. However, after the meiotic prophase, the oocyte does not immediately proceed to the first meiotic division, but enters a several-month-long growth phase. It grows up to a volume of about 1 μL, with a protein content of 25 μg, and then it stops. At this point, the cell is technically still in meiotic prophase, since transcription is taking place and the M-phase cyclins are present. However, these cyclins are locked in inactive complexes with CDK1, and thus the cell is arrested indefinitely in this state, with all its various opposing processes (protein synthesis/degradation, phosphorylation/dephosphorylation, anabolism/catabolism, etc.) in balance.

The meiosis process is resumed only when the ovarian epithelial cells surrounding the oocyte release a maturation-promoting hormone, progesterone, in response to gonadotropins produced by the frog pituitary. *Xenopus* oocytes possess both classical progesterone receptors and seven transmembrane G-protein-coupled progesterone receptors. However, progesterone undergoes metabolism in the oocyte, and there is evidence that androgens and androgen receptors may ultimately mediate progesterone's effects. Regardless of whether a progestin or an androgen is the ultimate trigger, the effects of progesterone on immature oocytes are striking. The oocyte leaves its G2-arrest state, carries out the first asymmetrical meiotic division, enters meiosis II and then arrests in the metaphase of meiosis II. This progression from the G2-arrest state to the meiosis II-arrest state is termed maturation. After maturation the oocyte is ovulated, acquires a jelly coat and is laid by the frog. It then drifts in the pond in this arrested state until either it is fertilised, which allows it to complete meiosis and commence embryogenesis, or it undergoes apoptosis.

Oocyte maturation is a typical example of a cell fate switch: the cell responds to an external trigger by undergoing an all-or-none, irreversible change in its appearance, its biochemical state and its developmental potential, [13].

Although many details of this system still remain to be elucidated, in broad outline the signalling network that mediates progesterone-induced oocyte maturation is well-understood and is depicted in Fig. 5.7. Progesterone stimulates the translation of the Mos oncoprotein, a MAP kinase kinase kinase (MAPKKK). Active Mos phosphorylates and activates the MAPKK MEK1, which then phosphorylates and activates ERK2 (which in *Xenopus* is often called p42 MAPK). Inhibitors of these MAPK cascade proteins inhibit oocyte maturation, and activated forms of the proteins can initiate maturation in the absence of progesterone. The activation of p42 MAPK then yields the dephosphorylation and activation of cyclin B-CDK1 complexes (sometimes named "latent MPF," for latent maturation-promoting factor or "pre-MPF"). Activated cyclin B-CDK1 complexes then cause the oocyte to resume the meiotic M-phase.

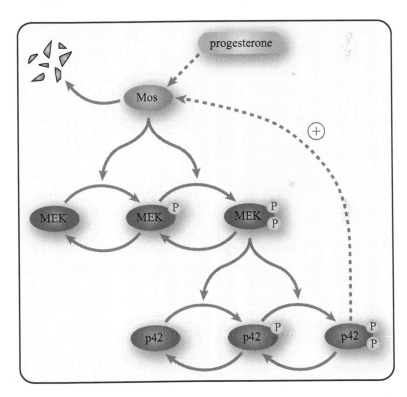

FIGURE 5.7: Schematic depiction of the Mos-MEK-p42 MAPK cascade. The system comprises a positive feedback loop consisting of active (double phosphorylated) p42 increasing the concentration of active Mos through a number of (not shown) intermediate steps.

In what follows, we will investigate the "all-or-nothing" character of oocyte maturation. In particular, we will use the concept of bistability and related analysis tools to understand how and under what conditions a network of reversible activation processes culminates in an irreversible cell fate change.

Note that the cascade is embedded in a positive feedback loop; indeed, the activation of p42 MAPK stimulates the accumulation of its upstream activator, the Mos oncoprotein, probably through both an increase in the rate of Mos translation and a decrease in the rate of Mos proteolysis. Thus, we will apply the Monotone Systems theory introduced in Section 5.3 to investigate the bistability of the Mos-MEK-p42 MAPK cascade, following the treatment in [5].

Breaking the positive feedback from p42 to Mos, we can write the open-loop model of the MAPK cascade as

$$\frac{dMos}{dt} = \frac{V_2 \cdot Mos}{K_2 + Mos} + V_0 \cdot \omega + V_1 \tag{5.15a}$$

$$\frac{dMEK}{dt} = \frac{V_6 \cdot MEK_p}{K_6 + MEK_p} - \frac{V_3 \cdot Mos \cdot MEK}{K_3 + MEK} \tag{5.15b}$$

$$\frac{dMEK_{pp}}{dt} = \frac{V_4 \cdot Mos \cdot MEK_p}{K_4 + MEK_p} - \frac{V_5 \cdot MEK_{pp}}{K_5 + MEK_{pp}} \tag{5.15c}$$

$$\frac{dp42}{dt} = \frac{V_{10} \cdot p42_p}{K_{10} + p42_p} - \frac{V_7 \cdot MEK_{pp} \cdot p42}{K_7 + p42} \tag{5.15d}$$

$$\frac{dp42_{pp}}{dt} = \frac{V_8 \cdot MEK_{pp} \cdot p42_p}{K_8 + p42_p} - \frac{V_9 \cdot p42_{pp}}{K_9 + p42_{pp}} \tag{5.15e}$$

$$MEK_p = MEK_{\text{tot}} - MEK - MEK_{pp} \tag{5.15f}$$

$$p42_p = p42_{\text{tot}} - p42 - p42_{pp} \tag{5.15g}$$

where ω is the input to the system and $p42_{pp} = \eta$ is the output. We have assumed that the total concentrations of MEK and p42 are constant, that is $MEK + MEK_p + MEK_{pp} = MEK_{\text{tot}}$ and $p42 + p42_p + p42_{pp} = p42_{\text{tot}}$. Therefore, the two differential equations for MEK_p and $p42_p$ have been substituted by the two algebraic conservation equations (5.15f) and (5.15g). The parameter values for these equations are shown in Table 5.1. In order to keep the analysis simple, we can easily decompose the MAPK cascade into three submodules, consisting of the three kinase levels:

I) MAPKKK module, consisting of just Mos, with input ω and output Mos;

II) MAPKK module, made up by MEK, MEK_p and MEK_{pp}, with input Mos and output MEK_{pp};

III) MAPK module, made up by p42, $p42_p$ and $p42_{pp}$, with input MEK_{pp} and output $p42_{pp}$.

TABLE 5.1

Parameters for model (5.15). The values have been chosen in [5] such that the model kinetics are consistent with experimentally available data

Parameter	Value	Unit	Parameter	Value	Unit
MEK_{tot}	1200	nM	$p42_{tot}$	300	nM
V_0	0.0015	$s^{-1} \cdot nM^{-1}$	V_1	2e−06	s^{-1}
V_2	1.2	$nM \cdot s^{-1}$	K_2	200	nM
V_2	1.2	$nM \cdot s^{-1}$	K_2	200	nM
V_3	0.064	s^{-1}	K_3	1200	nM
V_4	0.064	s^{-1}	K_4	1200	nM
V_5	5	$nM \cdot s^{-1}$	K_5	1200	nM
V_6	5	$nM \cdot s^{-1}$	K_6	1200	nM
V_7	0.06	s^{-1}	K_7	300	nM
V_8	0.06	s^{-1}	K_8	300	nM
V_9	5	$nM \cdot s^{-1}$	K_9	300	nM
V_{10}	5	$nM \cdot s^{-1}$	K_{10}	300	nM

Recall from Section 5.3 that, for each of the three modules, we have to verify that (A) the open-loop subsystem has a monostable steady-state response to constant inputs (also referred to as a well-defined steady-state I/O characteristic) and that (B) there are no possible negative feedback loops, even when the system is closed under positive feedback, which means the subsystem is strongly I/O monotone. Exploiting the modularity of the system, we can state that the whole system verifies properties (A) and (B) if they are satisfied by all of the three modules.

The satisfaction of property (A) can be verified by simulation, as shown in Fig. 5.8, where the steady-state I/O characteristics of the three submodules are depicted.

To check whether property (B) is also verified, we have to build the signed incidence graphs of the three modules (see Fig. 5.9). By visual inspection, it is straightforward to see that there are no negative feedback loops in the three graphs. Since the whole open-loop system is a cascade of these three modules, then also the whole graph will not contain any negative feedback loop.

Now that properties (A) and (B) have been checked, we can investigate the bistability of the MAPK cascade by drawing the steady-state I/O characteristic of the whole system, reported in Fig. 5.10. The diagram shows that there are two asymptotically stable equilibrium points, one at zero and one

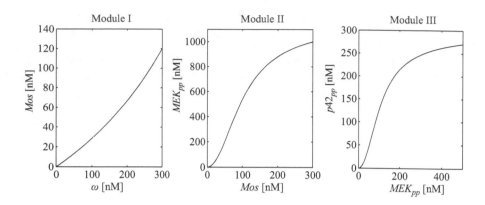

FIGURE 5.8: Steady-state I/O characteristics of the three submodules of the MAPK cascade. The diagrams show that the three subsystems all have a well-defined steady-state I/O characteristic.

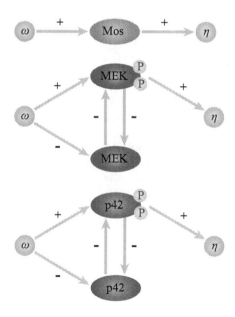

FIGURE 5.9: Signed incidence graphs of the three submodules of the MAPK cascade. We have indicated with ω and η the input and output of each module, respectively. The graphs show that the three subsystems have no negative feedback.

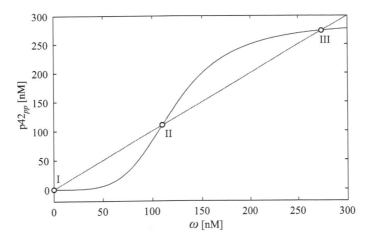

FIGURE 5.10: Steady-state I/O characteristic of the open-loop MAPK cascade model (5.15). The intersections with the line $\omega = \eta$ identify the equilibrium points: I and III are asymptotically stable, since the slope of the I/O characteristic is less than one, whereas II is unstable.

at a high concentration of $p42_{pp}$ and an intermediate unstable equilibrium; thus the system is bistable. Confirmation of this fact is provided in Fig. 5.11, which shows the time courses of the free evolution of the closed-loop system, starting from different initial concentrations of the kinases: the trajectories funnel into one or another of the two stable states, depending on the initial condition. Finally, in Fig. 5.12 a bifurcation diagram is used to show which values of the feedback gain parameter ν give rise to bistability: the diagram confirms that bistability occurs only for values of ν over a certain threshold. Moreover, we can see that the two stable steady-states (and the middle unstable one) coexist even for large values of ν, which is in agreement with what can be derived from Fig. 5.10.

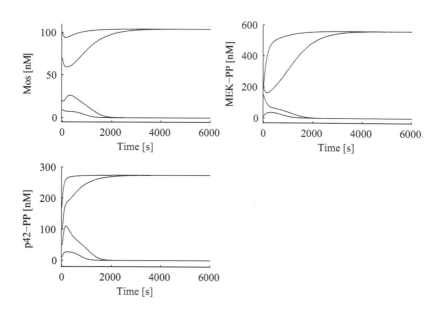

FIGURE 5.11: Evolution of the closed-loop MAPK cascade model starting from different random initial conditions. The trajectories converge to one of the two stable equilibrium points.

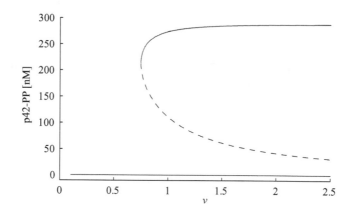

FIGURE 5.12: Bifurcation diagram of the closed-loop MAPK model, showing the steady-state concentrations of p42-PP as a function of the feedback gain ν.

5.6 Case Study VIII: Coupled positive and negative feedback loops in the yeast galactose pathway

Biology background: The capability to adapt to changing environmental conditions is a key evolutionary pressure in all living organisms. One of the primary needs of single-celled organisms such as yeasts is to adapt to constantly changing sources of nutrients, according to their availability in the surrounding environment. *Saccharomyces cerevisiae* has evolved an elaborate biomolecular circuit to control the expression of galactose-metabolising enzymes, in order to use galactose as an alternative carbon source in the absence of glucose.

This system consists of two positive and one negative (repressing) feedback loops, which affect the uptake of galactose, the nucleoplasmic shuttling of regulator proteins and the transcription of *GAL* genes. The *GAL* gene family in *S. cerevisiae* consists of three regulatory (*GAL4*, *GAL80* and *GAL3*) and five structural genes (*GAL1*, *GAL2*, *GAL7*, *GAL10* and *MEL1*), which enable it to use galactose as a carbon source. The structural genes *GAL1*, *GAL7* and *GAL10* are clustered but separately transcribed from individual promoters.

The regulatory network of the yeast galactose pathway is depicted in Fig. 5.13: gene *GAL4* encodes a transcriptional activator Gal4p that binds to the upstream activation sequences of *GAL* genes as a homodimer and activates the transcription of the genes. The repressor protein, Gal80p, self-associates to form a dimer and subsequently binds to the gene-Gal4p dimer complex and prevents it from recruiting RNA polymerase II mediator complex, thereby preventing the activation of *GAL* genes.

In the presence of inducer, galactose and adenosine triphosphate, Gal3p is activated and forms a complex with Gal80p in the cytoplasm. Binding of Gal3p affects the shuttling of Gal80p between the cytoplasm and the nucleus, reducing the concentration of Gal80p in the nucleus and, thus, relieving its inactivating effect on Gal4p and on the transcription of *GAL* genes. The transcription and translation of Gal2 produces the permease Gal2p, which mediates the transport of galactose into the cells. The increase of internalised galactose, in turn, further activates Gal3p. In the presence of glucose, on the other hand, the synthesis of Gal4p is inhibited through Mig1p-mediated repression of *GAL* genes, [14].

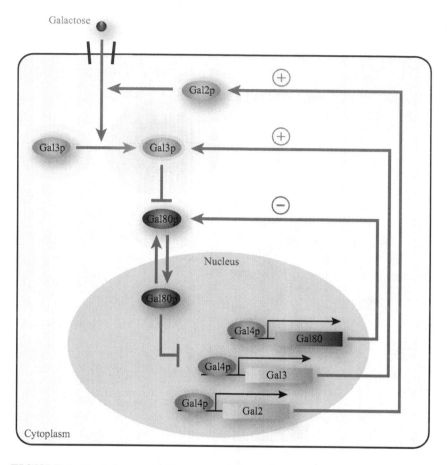

FIGURE 5.13: Schematic diagram of the galactose signalling pathway, highlighting the coupled positive and negative feedback loops.

Due to the presence of two feedback loops, the *GAL* regulatory network has the potential for exhibiting multistability. This capability has been evidenced experimentally [15] by growing wild-type cells for 12 hours either in the absence of galactose or in the presence of 2% galactose. In the absence of galactose, raffinose was used as a carbon source that does not induce or repress the *GAL* regulatory network. Subsequently, the cells were grown for a further 27 hours at various concentrations of galactose. It was observed that the responses of the two groups depend strongly on the galactose concentration. At low and high galactose concentrations the expression distributions after 27 hours do not depend on the previous treatment and they typically reach a steady-state after 6 hours. This behaviour is classified as history independent (absence of memory), because the system approaches the same unique expression distribution independently of the initial concentration. However, for intermediate galactose concentrations the expression distributions of the two groups are significantly different and the system displays a memory of the initial galactose consumption state. This experiment reveals a persistent memory, because cells become stably locked into two different expression states for periods much longer than the history-independent system would need to reach steady-state.

Several different models have been presented in the literature to investigate the behaviour of the galactose pathway, comprising simplified reduced-order models, [16],[17], and more comprehensive models also including the metabolic subsystem [18]. Such models have been useful for obtaining a more thorough understanding of the multistable dynamics of the GAL regulatory system; however, they have been mostly exploited by numerical simulations, in order to validate the hypothesised mechanisms by comparison with experimental results. Here, we show how it is possible to approach the issue of multistability in the GAL system by means of CRNT, thus providing a sound theoretical validation of the proposed mathematical model not solely based on data/parameter fitting, but on the structural properties of the reaction network.

Recall that, in order to exploit the CRNT Deficiency One Algorithm, we must have a differential equation model which exhibits only mass action kinetics. Therefore, we have built a novel model of the GAL system, focusing only on the regulatory subnetwork illustrated in Fig. 5.13, including the species in Table 5.2. Note that the purpose of this model is to study the bistability feature of the known galactose reaction network rather than providing a detailed description of the kinetics. Therefore, a number of simplifying assumptions have been made:

a) the regulatory mechanisms that are activated in the presence of glucose are neglected;

b) only the G2-mediated uptake of galactose is considered (in reality there is also a G2-independent intrinsic transport mechanism);

c) the cytoplasm and nucleus are not treated as separate compartments; thus the shuttling is not modelled;

d) no binding/unbinding of G4 to/from DNA is modelled;

e) dimerisation of proteins is neglected;

The reaction diagram of the proposed model is as follows.

TABLE 5.2

State variables of model (5.16)

State variable	Description
G_2	Gal2 protein concentration
G_3	Gal3p protein concentration
G_4	Gal4p protein concentration
G_{80}	Gal80p protein concentration
G_{3a}	Active Gal3p protein concentration
$G_{4,80}$	Gal4p:Gal80p complex concentration
$G_{3a,80}$	Active Gal3:Gal80p complex concentration
G_i	Internalised galactose concentration
G_e	Extracellular galactose concentration

$$G_4 \rightleftarrows \varnothing \qquad\qquad G_i \longrightarrow \varnothing \qquad\qquad G_3 + G_i \rightleftarrows G_{3a} \longrightarrow \varnothing$$

$$G_3 + G_4 \qquad G_2 + G_4 \qquad G_2 \longrightarrow \varnothing \qquad\qquad G_{3a,80} \longrightarrow \varnothing$$

$$G_4 + G_{80} \rightleftarrows G_{4,80} \longrightarrow \varnothing \qquad G_3 \longrightarrow \varnothing \qquad G_{4,80} + G_{3a} \rightleftarrows G_{3a,80} + G_4$$

$$G_{80} \longrightarrow \varnothing \qquad G_e + G_2 \longrightarrow G_i + G_2 + G_e$$

where $\varnothing$ denotes the null species, which allows us to model protein degradation and generation. Note that, to model an extracellular medium with a constant concentration of galactose, the reaction describing G_2-mediated uptake of galactose,

$$G_e + G_2 \rightarrow G_i + G_2 + G_e,$$

creates a new molecule of external galactose G_e for every internalised molecule G_i. Induction of transcription/translation of G_2, G_3, G_{80} is modelled by simple reactions of the type $G_4 \rightarrow G_x + G_4$, where G_4 is both a reagent and a product since it is not modified in the process. The inactivation of the inhibitor is synthetically described by the reaction

$$G_{4,80} + G_{3a} \rightleftarrows G_{3a,80} + G_4$$

which models only the binding of G_3 to those G_{80} molecules which are bound to the transcription factor G_4 and the subsequent release of the latter protein. Finally, the reaction

$$G_3 + G_i \rightleftarrows G_{3a}$$

describes the activation of G_3 by internalised galactose, assuming that the latter is consumed in the reaction. From the reaction diagram, assuming mass action kinetics for each reaction rate, it is easy to derive the dynamical model which describes the changes over time of the species concentrations. The model is given by

$$\dot{G}_3 = k_9\, G_4 - k_1\, G_3\, G_i + k_2 G_{3a} - \mu_1 G_3 \tag{5.16a}$$

$$\dot{G}_i = k_{11}\, G_e\, G_2 - \mu_8\, G_i - k_1\, G_3\, G_i + k_2\, G_{3a} \tag{5.16b}$$

$$\dot{G}_{3a} = k_1\, G_3\, G_i - k_2\, G_{3a} - \mu_3\, G_{3a} - k_5\, G_{4,80}\, G_{3a} + k_6\, G_{3a,80}\, G_4 \tag{5.16c}$$

$$\dot{G}_4 = k_{10} - \mu_4\, G_4 + k_5\, G_{4,80}\, G_{3a} - k_6\, G_{3a,80}\, G_4 - k_3\, G_4\, G_{80} + k_4\, G_{4,80} \tag{5.16d}$$

$$\dot{G}_{80} = -\mu_2\, G_{80} - k_3\, G_4\, G_{80} + k_4\, G_{4,80} + k_7\, G_4 \tag{5.16e}$$

$$\dot{G}_{4,80} = k_3\, G_4\, G_{80} - k_4\, G_{4,80} - \mu_6\, G_{4,80} - k_5\, G_{4,80}\, G_3 a + k_6\, G_{3a,80}\, G_4 \tag{5.16f}$$

$$\dot{G}_{3a,80} = k_5\, G_{4,80}\, G_{3a} - k_6\, G_{3a,80}\, G_4 - \mu_7\, G_{3a,80} \tag{5.16g}$$

$$\dot{G}_2 = k_8\, G_4 - \mu_5\, G_2 \tag{5.16h}$$

$$\dot{G}_e = 0 \tag{5.16i}$$

At this point we apply the CRNT toolbox to determine whether system (5.16) can admit multiple steady-states. After introducing the species and the reactions, the toolbox returns a *basic report*, which informs us about the graphical properties of the network: there are seventeen complexes, fifteen reactions and three linkage classes (note that all the reactions including the null species form a single linkage class, although we have drawn them separately for clarity). The software also informs us that there are four terminal strong linkage classes and that the network is neither reversible nor weakly reversible. The rank of the network is eight, the deficiencies of the three linkage classes are four, zero and zero, respectively, while the whole network has deficiency six. Hence, the basic theorems introduced in Section 5.4 cannot establish whether the network is bistable; however, the report states that further analyses can be conducted using some extensions of the theory, namely the Mass Action Injectivity analysis, [19], and Higher Deficiency analysis, [20]. In particular, from the latter analysis the network is proved to have the capacity for multiple steady-states, and the software also provides an example set of rate constants for which two steady-states (which are reported as well) exist. The values of the kinetic parameters and of the two steady-states are shown in Tables 5.3 and 5.4. These values are found by means of an optimisation procedure without reference to any experimental measurement; therefore they are assigned arbitrary units. Moreover, they cannot be considered as valid measures of biological kinetic parameters, because there is no guarantee that this is the only combination of values that results in bistability of the model. Nevertheless, we can gain further insight into the system's basic mechanisms by examining these values. For example, note that there are some quantities which do not change significantly between the two equilibrium points, while others exhibit large changes.

TABLE 5.3
Kinetic parameters values (arbitrary units)
which make model (5.16) bistable

Parameter	Value	Parameter	Value
k_1	7.353E-3	k_2	7.078
k_3	28.28	k_4	0.1158
k_5	12.03	k_6	3.741
k_7	31.67	k_8	1
k_9	86.79	k_{10}	9.639
k_{11}	86.79	μ_1	1
μ_2	1	μ_3	1
μ_4	1	μ_5	1
μ_6	1	μ_7	1
μ_8	1		

TABLE 5.4
Species concentrations (arbitrary units) at steady-state
equilibrium points for the bistable model (5.16) with parameter
values given in Table 5.3

State variable	Value at equilibrium 1	Value at equilibrium 2
G_3	63.87	105.3
G_i	63.87	105.3
G_{3a}	1	4.056
G_4	1	1.822
G_{80}	1.116	1.116
$G_{4,80}$	8.639	7.817
$G_{3a,80}$	21.92	48.78
G_e	1	1
G_2	1	1.822

This could lead us to conclude that changes in the concentrations of the latter
species correspond to those that play the largest role in determining the final
steady-state.

In Fig. 5.14, we show the response of system (5.16) to different initial con-

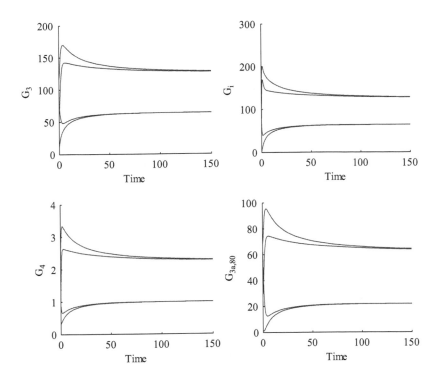

FIGURE 5.14: Free evolutions, for different initial conditions, of the concentrations of four species of the galactose regulatory network model (5.16) with parameter values given in Table 5.3. The curves funnel into either one of two steady-states, confirming the bistable nature of the system.

ditions: the plots confirm the bistable behaviour of the proposed galactose model. In particular, when at time zero G_3, G_i, G_4 and $G_{3a,80}$ are low, the system reaches the low equilibrium value, while the high equilibrium value is reached by imposing large initial concentrations. These simulations resemble the experiments in which the cells have been precultured without and with galactose, respectively. Indeed, pre-culturing the cells in the absence (resp. in the presence) of galactose leads to a down-regulation (resp. an up-regulation) of the *GAL* genes, that is initial low (resp. high) values of $G2$, $G3$ and $G4$ in the subsequent experimental phase.

References

[1] Maayan A, Iyengar R, and Sontag ED. Intracellular regulatory networks are close to monotone systems. *IET Systems Biology*, 2:103–112, 2008.

[2] Strogatz SH. *Nonlinear Dynamics and Chaos*. Reading: Perseus Books Publishing, 1994.

[3] Hanusse P. De l'éxistence d'un cycle limit dans l'évolution des systèmes chimique ouverts (on the existence of a limit cycle in the evolution of open chemical systems). *Comptes Rendus, Acad. Sci. Paris*, (C), 274:1245–1247, 1972.

[4] Schnakenberg J. Simple chemical reaction systems with limit cycle behaviour. *Journal of Theoretical Biology*, 81(3):389–400, 1979.

[5] Angeli D, Ferrell JE, and Sontag ED. Detection of multistability, bifurcations, and hysteresis in a large class of biological positive-feedback systems. *PNAS*, 101(7):1822–1827, 2004.

[6] Angeli D and Sontag ED. Multistability in monotone input/output systems. *Systems and Control Letters*, 51(3-4):185–202, 2004.

[7] Angeli D and Sontag ED. Monotone control systems. *IEEE Transactions on Automatic Control*, 48(10):1684–1698, 2003.

[8] Feinberg M. Chemical reaction network structure and the stability of complex isothermal reactors — I. The deficiency zero and deficiency one theorems. *Chemical Engineering Science*, 42(10):2229–2268, 1987.

[9] Feinberg M. Chemical reaction network structure and the stability of complex isothermal reactors — II. Multiple steady states for network of deficiency one. *Chemical Engineering Science*, 43(1):1–25, 1988.

[10] Conradi C, Saez-Rodriguez J, Gilles E-D, and Raisch J. Using chemical reaction network theory to discard a kinetic mechanism hypothesis. *IEEE Proceedings Systems Biology*, 152(4):243–248, 2005.

[11] Siegal-Gaskins D, Grotewold E, and Smith GD. The capacity for multistability in small gene regulatory networks. *BMC Systems Biology*, 3:96, 2009.

[12] Palsson BØ. *Systems Biology: Properties of Reconstructed Networks*. Cambridge: Cambridge University Press, 2006.

[13] Ferrell JE, Pomerening JR, Young Kim S, Trunnell NB, Xiong W, Frederick Huang C-Y, and Machleder EM. Simple, realistic models of complex biological processes: Positive feedback and bistability in a cell fate switch and a cell cycle oscillator. *FEBS Letters*, 583:3999–4005, 2009.

[14] Pannala VR, Bhat PJ, Bhartiya S, and Venkatesh KV. Systems biology of Gal regulon in Saccharomyces cerevisiae. *WIREs Systems Biology and Medicine*, 2:98–106, 2010.

[15] Acar M, Becskei A, and van Oudenaarden A. Enhancement of cellular memory by reducing stochastic transitions. *Nature*, 435:228–232, 2005.

[16] Smidtas S, Schächter V, and Képès F. The adaptive filter of the yeast galactose pathway. *Journal of Theoretical Biology*, 242:372–381, 2006.

[17] Kulkarni VV, Kareenhalli V, Malakar P, Pao LY, Safonov MG and Viswanathan GA. Stability analysis of the GAL regulatory network in *Saccharomyces cerevisiae* and *Kluyveromyces lactis*. *BMC Bioinformatics*, 11(Suppl 1):S43, 2010.

[18] de Atauri P, Orrell D, Ramsey S, and Bolouri H. Evolution of design principles in biochemical networks. *IEEE Proceedings Systems Biology*, 1:28–40, 2004.

[19] Craciun G and Feinberg M. Multiple equilibria in complex chemical reaction networks. I. The injectivity property. *SIAM Journal on Applied Mathematics*, 65:1526–1546, 2005.

[20] Ellison P. The advanced deficiency algorithm and its applications to mechanism discrimination. PhD. Thesis. Rochester, NY: Department of Chemical Engineering, University of Rochester, 1998.

6

Model validation using robustness analysis

6.1 Introduction

Robustness, the ability of a system to function correctly in the presence of both internal and external uncertainty, has emerged as a key organising principle in many biological systems. Biological robustness has thus become a major focus of research in systems biology, particularly on the engineering–biology interface, since the concept of robustness was first rigorously defined in the context of engineering control systems. This chapter focuses on one particularly important aspect of robustness in systems biology, i.e. the use of robustness analysis methods for the validation or invalidation of models of biological systems. With the explosive growth in quantitative modelling brought about by systems biology, the problem of validating, invalidating and discriminating between competing models of a biological system has become an increasingly important one. In this chapter, we provide an overview of the tools and methods which are available for this task, and illustrate the wide range of biological systems to which this approach has been successfully applied.

The case for robustness being a key organising principle of biological systems was first made in an influential series of papers in the early 2000's, [1, 2]. In these papers, the authors compare the robustness properties of biological and engineered systems, and suggest that the need for robustness is a key driver of complexity in both cases — radically simplified versions of both jet aircraft and bacteria could be conceived of that would function in highly controlled "laboratory" conditions, but would lack the robustness properties necessary to function correctly in highly fluctuating real-world environments. Somewhat paradoxically, the highly complex nature of these systems renders them "robust yet fragile," that is, robust to types of uncertainty or variation that are common or anticipated, but potentially highly fragile to rare or unanticipated events. For example, biological organisms are usually highly robust to uncertainty in their environments and component parts but can be catastrophically disabled by tiny perturbations to genes or the presence of microscopic pathogens or trace amounts of toxins that disrupt structural elements or regulatory control networks. Complex biological control systems such as the heat shock response result in highly robust performance but also generate new fragilities which must be compensated for by other systems, [3]. In a similar manner, modern high-performance aircraft are robust to large-scale atmospheric disturbances, variations in cargo loads and fuels, turbulent boundary layers, and inhomogeneities and aging of materials, but could be catastrophically disabled by microscopic alterations in a handful of very large-scale integrated chips or by software failures (in contrast to previous generations of much more simple "mechanical" aircraft which had little or no reliance on computers). This theme has since been developed to form the basis of a coherent theory of biological robustness, [4]–[9].

In this chapter, we focus on one of the most practically useful ideas which has emerged from this sometimes rather philosophical line of enquiry. This idea was first made explicit in [10], and is perfectly encapsulated in the title of the paper: Robustness as a measure of plausibility in models of biochemical networks. The idea is of course an entirely logical consequence of the recognition of the robust nature of biological systems: if a particular feature of a system has been shown experimentally to be robust to a certain kind of perturbation or environmental disturbance, then any proposed model of this system should also demonstrate the same levels of robustness to simulated versions of the same perturbations or disturbances. The great advantage of this idea is that it provides a much more stringent "test" of a proposed model than the traditional approach of simply asking: does there exist a biologically plausible set of model parameter values for which the model's outputs provide an acceptable match to experimental data?

As the complexity of the quantitative models being developed in systems biology research continues to escalate, it is obvious that it will often be the case that many, conceptually quite different, models may be proposed to "explain" the workings of a biological system, and that each of these models will often have biologically reasonable sets of parameter values which allow the model to accurately reproduce the experimentally measured dynamics of the system. Since each of these models encapsulates a different hypothesis regarding the workings of the underlying biology, it is clear that further progress depends on the ability to reliably discriminate between different models, discarding some and focussing on others for further refinement, development and testing.

Here, we use the term "model validation" to describe this process, although to be precise, as pointed out in [11], the complete validation of a particular model is never possible in practice, as it would require infinite amounts of both data and computational power. Usually, the best one can do is to proceed by a process of elimination, invalidating more and more competing models until a single uninvalidated model remains. This model then encapsulates our current level of understanding of the underlying biology, which may stand the test of time, or be subsequently refined in the light of new data. The evaluation of model robustness provides a powerful tool with which to achieve the goal of developing validated models of biological reality, and this approach has now been used as an essential part of the model development process for a wide range of biological systems, [12, 13, 14, 15, 16, 17].

6.2   Robustness analysis tools for model validation

In this section, we describe the tools and techniques which are available to evaluate the robustness of models of biological systems to various forms of uncertainty and variability. Many of these methods were first developed within the field of control engineering, where linear models, or models with particular forms of nonlinearity, are typically used for the purposes of design and analysis. Biological systems, on the other hand, often display highly complex behaviour, including strong nonlinearities, as well as oscillatory, time-varying, stochastic and/or hybrid discrete-continuous dynamics. Thus, the application of these methods in the context of systems biology is often far from straightforward, and care must often be exercised in interpreting the computed results. As shown below, however, careful analysis of systems biology models using these tools can often provide significant insight into both the validity of a particular model and the underlying biological mechanisms it represents.

6.2.1   Bifurcation diagrams

Biological systems typically operate in the neighbourhood of some nominal condition, e.g. in biochemical networks the production and degradation rates of the biochemical compounds are often regulated so that the amounts of each species remain approximately constant at some levels. When such an *equilibrium* is perturbed by an unpredicted event (e.g. by the presence of exogenous signalling molecules, like growth factors), a variety of different reactions may take place, which in general can lead the system either to operate at a different equilibrium point, or to tackle the cause of the perturbation in order to restore the nominal operative condition.

Since, in nonlinear systems, the equilibrium points of a system and their stability properties depend not just on the structure of the equations but also on the values of the parameters, even small changes in the value of a single parameter can significantly alter the map of equilibrium points, and thus the dynamic behaviour of the system: this phenomenon is called a *bifurcation*. As described in Section 5.2, the variations in the map of equilibrium points corresponding to changes in one or more model parameters can be effectively visualised by using a *bifurcation diagram*, in which the equilibrium values of some state variable are plotted against the bifurcation parameter.

Bifurcation diagrams are powerful tools for understanding how qualitative changes in the behaviour of nonlinear systems biology models arise due to parametric uncertainty. As tools for measuring robustness, however, they suffer from two significant limitations, namely, that analytical solutions are available only for low-order models, and that they only provide information

on the effects of varying one or two parameters at a time.* Nonetheless, bifurcation analysis was the tool used in the first paper proposing the use of
robustness analysis for model validation: in [10], a model of the biochemical
oscillator underlying the *Xenopus* cell cycle was represented as a mapping
from parameter space to behaviour space, and bifurcation analysis was used
to study the robustness of each region of steady-state behavior to parameter variations. The hypothesis that potential errors in models will result in
parameter sensitivities was tested by analysis of the robustness of two different models of the biochemical oscillator. This analysis successfully identified
known weaknesses in an older model and also correctly highlighted why the
more recent model was more plausible. In [18], a bifurcation analysis software
package named AUTO was employed to examine the robustness of a model of
cAMP oscillations in aggregating *Dictyostelium* cells to variations in each of
the kinetic constants k_i in the model, while in [19], the authors use bifurcation analysis to compare the validity of high- and low-order models describing
regulation of the cyclin-dependent kinase that triggers DNA synthesis and
mitosis in yeast. Finally, in [20], the authors introduce a novel robustness
analysis method for oscillatory models, based on the combination of Hopf bifurcation analysis and the standard Routh–Hurwitz stability test from linear
control theory.

6.2.2   Sensitivity analysis

Sensitivity analysis is a well-established technique for evaluating the relative
sensitivity of the states or outputs of a model to changes in its parameters. In
this sense, therefore, sensitivity may be interpreted as the inverse of robustness — parameter sensitivities yield a quantitative measure of the deviations
in characteristic system properties resulting from perturbation of system parameters and thus a higher (absolute) sensitivity of a parameter implies a
lower robustness of the corresponding element of a model. The classical approach to sensitivity analysis considers small variations in a single parameter
at a time. For the autonomous dynamical system described by the ordinary
differential equation

$$\dot{x} = f\left(x(t), p, t\right) \tag{6.1}$$

with time $t \geq t_0$, the $n_S \times 1$ vector of state variables x, the $n_P \times 1$ vector of
model parameters p and initial conditions $x(t_0) = x_0$, parameter sensitivities
with respect to the system's states along a specific trajectory $S(t)$ (the $n_S \times n_P$

---

*In principle, one could consider more parameters but the dynamic behaviour near bi-
furcations with codimension higher than three is usually so poorly understood that the
computation of such points is not worthwhile.

matrix of state sensitivities) are defined by[†]

$$S(t) = \frac{\delta x}{\delta p} \tag{6.2}$$

To allow for easier comparisons to be made between different models, the sensitivity of each parameter p_j may be integrated over discrete time points along the system's trajectory from T_0 to T_{n_T}, and normalised to relative sensitivity (log-gain sensitivity) to give the overall state sensitivity for parameter p_j:

$$S_{Oj}(t) = \frac{1}{n_S} p_j \left(\sum_{k=1}^{n_T} \sum_{i=1}^{n_S} \left[\frac{1}{x_i} \frac{\delta x_i(t_k, t_0)}{\delta p_j} \right]^2 \right)^{1/2} \tag{6.3}$$

The sensitivity of each parameter with respect to any model output, or other characteristic, may be evaluated in the same way; for example, the sensitivity of the period and amplitude of an oscillatory system are evaluated, respectively, as

$$S_\tau = \frac{\delta \tau}{\delta p}, \text{ and } S_{A_i} = \frac{\delta A_i}{\delta p}. \tag{6.4}$$

It is important to note that the above parameter sensitivities are only valid locally with respect to a particular point in the model's parameter space, that is, in a neighbourhood of a specific parameter set. They thus only provide information on the robustness of a particular parameterisation of a model, and care must be taken in interpreting their values globally.

To derive global measures of parametric sensitivity, [21], some kind of gridding or sampling strategy must be used, in order to evaluate the relative sensitivity of different parameters over the full range of their allowable values. Of course, this significantly increases the associated computational cost, and also makes the direct comparison of the sensitivity of different parameters more difficult (relative sensitivities may vary across different regions of parameter space).

Nevertheless, in [22], the above sensitivity metrics were successfully used to investigate the specific structural characteristics that are responsible for robust performance in the genetic oscillator responsible for generating circadian rhythms in *Drosophila*. By systematically evaluating local sensitivities throughout the model's parameter space, global robustness properties linked to network structure could be derived. In particular, analysis of two mathematical models of moderate complexity showed that the tradeoff between robustness and fragility was largely determined by the regulatory structure. An analysis of rank-ordered sensitivities allowed the correct identification of protein phosphorylation as an influential process determining the oscillator's period. Furthermore, sensitivity analysis confirmed the theoretical insight

---

[†]Of course, analytical expressions for the relevant derivatives will rarely be available and thus numerical approximations will typically have to be employed.

that hierarchical control might be important for achieving robustness. The complex feedback structures encountered *in vivo* were shown to confer robust precision and adjustability of the clock while avoiding catastrophic failure.

Two recent papers have proposed effective strategies for overcoming the local, one-parameter-at-a-time limitations of traditional sensitivity analysis. In [23], the authors used sensitivity analysis to validate a new computational model of signal transducer and activator of transcription-3 (Stat3) pathway kinetics, a signaling network involved in embryonic stem cell self-renewal. Transient pathway behaviour was simulated for a 40-fold range of values for each model parameter in order to generate Stat3 activation surfaces — by examining these surfaces for local minima and maxima, non-monotonic effects of individual parameters could be identified and isolated. This analysis provided a range of parameter variations over which Stat3 activation is monotonic, thus facilitating a global sensitivity analysis of parameter interactions. To do this, groups of parameters which had a similar impact on pathway output were clustered together, so that the effects of varying multiple parameters at a time could be analysed visually using a clustergram.

This analysis allowed the identification of groups of parameters that contribute to pathway activation or inhibition, as well as other interesting pathway interactions. For example, it was found that simultaneously changing the parameters determining the nuclear export rate of Stat3 and the rate of docking of Stat3 on activated receptors influenced Stat3 activation more significantly than either of these parameters in isolation or in combination with any other parameters. It was further demonstrated that nuclear phosphatase activity, inhibition of SOCS3 and Stat3 nuclear export most significantly influenced Stat3 activation. These results were unaffected by how much parameters were changed, and could be averaged over different fold-changes in parameter values. The results of the sensitivity analysis were experimentally validated by using chemical inhibitors to specifically target different pathway activation steps and comparing the effects on the resultant Stat3 activation profiles with model predictions.

A different approach was adopted in [24], to produce what the authors refer to as a "glocal" robustness analysis (see Fig. 6.1) of two competing models of the cyanobacterial circadian oscillator. This two stage approach begins by sampling a large set of parameter combinations spanning several orders of magnitude for each parameter. From this sampling a subset of "viable" parameter combinations is selected which preserves the particular performance features of interest. Further sampling is conducted via an iterative scheme, where in each step the sampling distribution is adjusted based on a Principle Component Analysis (PCA) of the viable set of the previous step. After a Monte Carlo integration, the volume occupied by the set provides a first, crude characterisation of a model's robustness and can aid in model discrimination by proper normalisation. The second stage of the proposed approach defines a set of appropriate normalised local robustness metrics, e.g. a measure of how fast the oscillator returns to its cycling behaviour when its trajectory is

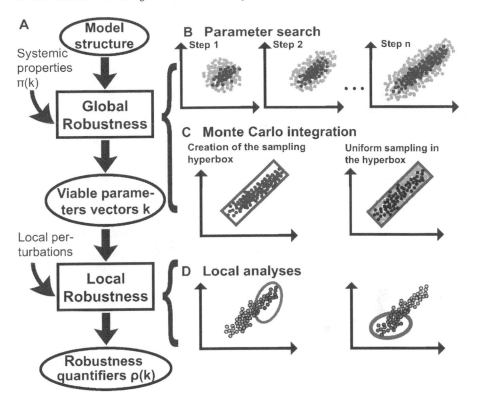

FIGURE 6.1: "Glocal" robustness analysis method, [24].

transiently perturbed with the use of Floquet multipliers, or the sensitivity of the period to perturbations in individual parameters or parameter vectors. These metrics are then evaluated for each viable parameter combination identified in the previous stage, and statistical tests are used to assess the analysis results.

Using this approach, two models based on fundamentally different assumptions about the underlying mechanism of the cyanobacterial circadian oscillator, termed the *autocatalytic* and *two (phosphorylation) sites* models, respectively, were compared in [24]. The results of this analysis showed that the *two sites* model had significantly better global and overall local robustness properties than the other model, hence making the assumptions on which it is based a more plausible explanation of the underlying biological reality.

6.2.3 μ-analysis

In this section, we describe a tool for measuring the robustness of a model to *simultaneous* variations in the values of several of its parameters. Since its introduction in the early days of robust control theory, [25, 26, 27], the structured singular value or μ has become the tool of choice among control engineers for the robustness analysis of complex uncertain systems.

It is generally possible to arrange any linear time invariant (LTI) system which is subject to some type of norm-bounded uncertainty in the form shown in Fig. 6.2, where M represents the known part of the system and Δ represents the uncertainty present in the system. Partitioning M compatibly with the Δ

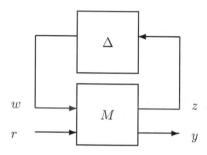

FIGURE 6.2: Upper LFT uncertainty description.

matrix, the relationship between the input and output signals of the closed-loop system shown in Fig. 6.2 is then given by the upper linear fractional transformation (LFT):

$$y \;=\; \mathcal{F}_u(M,\Delta)\, r \;=\; (M_{22} + M_{21}\Delta(I - M_{11}\Delta)^{-1}M_{12})\, r \qquad (6.5)$$

Now, assuming that the nominal system M in Fig. 6.2 is asymptotically stable and that Δ is a complex unstructured uncertainty matrix, the Small Gain Theorem (SGT), [27], gives the following result:
The closed-loop system in Fig. 6.2 is stable if

$$\bar{\sigma}(\Delta(j\omega)) \;<\; \frac{1}{\bar{\sigma}(M_{11}(j\omega))} \quad \forall\, \omega \qquad (6.6)$$

where $\bar{\sigma}$ denotes the maximum singular value. The above result defines a test for stability (and thus a robustness measure) for a system subject to *unstructured uncertainty* in terms of the maximum *singular value* of the matrix M_{11}.

Now, in cases where the uncertainty in the system arises due to variations in specific parameters, the uncertainty matrix Δ will have a diagonal or block diagonal structure, i.e.,

$$\Delta(j\omega) = diag(\Delta_1(j\omega), ..., \Delta_n(j\omega)), \ \overline{\sigma}(\Delta_i(j\omega)) \leq k \ \forall \ \omega \qquad (6.7)$$

Now again assume that the nominal closed-loop system is stable, and consider the question: What is the maximum value of k for which the closed-loop system will remain stable? We can still apply the SGT to the above problem, but the result will be conservative, since the block diagonal structure of the matrix Δ will not be taken into account. The SGT will in effect assume that all of the elements of the matrix Δ are allowed to be non-zero, when we know that most of the elements are in fact zero. Thus the SGT will consider a larger set of uncertainty than is in fact possible, and the resulting robustness measure will be conservative, i.e. pessimistic.

In order to get a non-conservative solution to this problem, Doyle, [25], introduced the structured singular value μ:

$$\mu_\Delta(M_{11}) = \frac{1}{\min(k \ \text{s.t.} \ \det(I - M_{11}\Delta) = 0)} \qquad (6.8)$$

The above result defines a test for stability (robustness measure) of a closed-loop system subject to *structured uncertainty* in terms of the maximum *structured singular value* of the matrix M_{11}. Singular value performance requirements can also be combined with stability robustness analysis in the μ framework to measure the *robust performance* properties of the system.

An obvious limitation of the μ framework is that it can only be applied to linear systems and thus only provides local robustness guarantees about an equilibrium. A second complicating factor is that the computation of μ is an NP hard problem, i.e. the computational burden of the algorithms that compute the exact value of μ is an exponential function of the size of the problem. It is consequently impossible to compute the exact value of μ for large dimensional problems, but an effective solution in this case is to compute upper and lower bounds on μ, and efficient routines for μ-bound computation are now widely available, [28]. Note that to fully exploit the power of the structured singular value theory, tight upper *and* lower bounds on μ are required. The upper bound provides a sufficient condition for stability/performance in the presence of a specified level of structured uncertainty. The lower bound provides a sufficient condition for *instability*, and also returns a worst-case Δ, i.e. a worst-case combination of uncertain parameters for the problem. The degree of difficulty involved in computing good bounds on μ depends on (a) the order of the Δ matrix, and (b) whether Δ is complex, real or mixed; see [28] for a full discussion.

In [18], μ-analysis was employed to evaluate the robustness of a biochemical network model which had been proposed to explain the capability of aggregating *Dictyostelium* cells to produce stable oscillations in the concentrations of intra- and extra-cellular cAMP. Due to the large number of uncertain parameters in the model, standard routines for computing lower bounds on μ

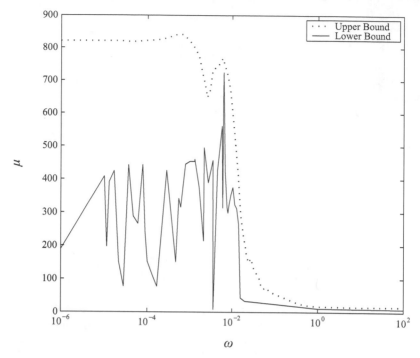

FIGURE 6.3: μ bounds for *Dictyostelium* network robustness analysis, [29].

failed for this problem, so that only an upper bound could be computed. Interestingly, and in contrast to the results of a parameter-at-a-time sensitivity analysis, this upper bound suggested a possible high degree of fragility in the model. This lack of robustness was subsequently confirmed by further analyses using a newly developed μ lower bound algorithm [29]. As shown in Fig. 6.3, simultaneous perturbations in the model's kinetic parameters of $1/723 = 0.14\%$ are sufficient to destabilise the oscillations, in stark contrast to the original claims that variations in model parameters over several orders of magnitude had little effect on its dynamics.

μ-analysis was also successfully employed in [30, 31, 32] to investigate the structural basis of robustness in the mammalian circadian clock. Systematic perturbations in the model structure were introduced, and the effects on the functionality of the model were quantified using the peak value of μ. Although in principle only one feedback loop involving the *Per* gene is required in the chosen clock model to generate oscillations, analysis using the structured singular value revealed that the presence of additional feedback loops involving the *Bmal1* and *Cry* genes significantly increases the robustness of the regulatory network. In [33], a similar approach was also used to validate models of oscillatory metabolism in activated neutrophils.

6.2.4 Optimisation-based robustness analysis

In robustness analysis, numerical optimisation algorithms can be used to search for particular combinations of parameters in the model's parameter space that maximise the deviation of the model's dynamic behaviour from experimental observations over a certain simulation time period. This type of search can be formulated as an optimisation problem of the form

$$\max_{p} c(x, p) \text{ subject to } \underline{p} \leq p \leq \overline{p} \tag{6.9}$$

where x is a vector of model parameters with upper and lower bounds $\overline{p}$ and $\underline{p}$, respectively, and $c(x, p)$ is an *objective function* or *cost function* representing the difference between the simulated outputs of the model and one or more sets of corresponding experimental data. By systematically varying the allowed level of uncertainty (defined by $\underline{p}$ and $\overline{p}$) in the model's parameters, and using the optimisation algorithm to compute the values of the model parameters which maximise this function, an accurate assessment of the model's robustness can be derived. A particular advantage of this approach is that it places little or no constraints on the form or complexity of the model — as long as it can be simulated with reasonable computational overheads, no additional modelling or analytical work is required to apply this approach. This is in sharp contrast to certain analytical approaches, such as μ-analysis or Sum-of-Squares programming (see below), which require the model to be represented in a particular form before any analysis can be conducted.

Due to the complex dynamics and large number of uncertain parameters in many systems biology models, the optimisation problems arising in the context of robustness analysis will generally be non-convex, and thus local optimisation methods, which can easily get locked into local optima in the case of multimodal search spaces, are often of limited use. Global optimisation methods, whether based on evolutionary principles, [34], or deterministic heuristics, [35], are usually much more effective, especially when coupled with local gradient-based algorithms via a hybrid switching strategy, [36]. This was the approach adopted in [37], where numerical optimisation algorithms were applied directly to a nonlinear biochemical network model to confirm an apparent lack of robustness indicated by a linear analysis using the structured singular value. Interestingly, it appears that the idea of using global optimisation to analyse the robustness and validity of complex simulation models was not first proposed in an engineering context, but by social scientists, who labeled the technique "Active Nonlinear Tests (ANTs)," [38]. Optimisation-based approaches have also recently been successfully applied to validate medical physiology simulation models in [39].

6.2.5   Sum-of-squares polynomials

Sum-of-Squares (SOS) programming has recently been introduced in the systems biology literature as a powerful new framework for the analysis and validation of a wide class of models, including those with nonlinear, continuous, discrete and hybrid dynamics, [40, 41]. A polynomial $p(y)$, with real coefficients, where $y \in R^n$, admits an SOS decomposition if there exist other polynomials $q_1, ..., q_m$ such that

$$p(y) = \sum_{i=1}^{m} q_i^2(y) \qquad (6.10)$$

where the subscripts denote the index of the m polynomials. If $p(y)$ is SOS, it can be easily seen that $p(y) \geq 0$ for all y, which means that $p(y)$ is non-negative. Polynomial non-negativity is a very important property (as many problems in optimisation and systems theory can be reduced to it) which is, however, very difficult to test (it has been shown to be NP-hard for polynomials of degree greater than or equal to 4). The existence of an SOS decomposition is a powerful relaxation for non-negativity because it can be verified in polynomial time. The reason for this, [42], is that $p(y)$ being SOS is equivalent to the existence of a positive semidefinite matrix Q (i.e. Q is symmetric and with non-negative eigenvalues) and a chosen vector of monomials $Z(y)$ such that

$$p(y) = Z^T(y)QZ(y) \qquad (6.11)$$

This means that that the SOS decomposition of $p(y)$ can be efficiently computed using Semidefinite Programming, and software capable of formulating and solving these types of problems is now widely available, [41]. To see how this framework can be applied to the problem of model validation (or more precisely, model *invalidation*), consider a model in the form of an autonomous, ordinary differential equation (ODE)

$$\dot{x} = f(x, p) \qquad (6.12)$$

where p is a vector in the allowable set of parameters $\mathcal{P}$ for the model and f satisfies appropriate smoothness conditions in order to ensure that given an initial condition there exists a locally unique solution. Now, for the system in question, assume that a set of experimental data $(t_i, \hat{x}_i)$ for $i = 1, ..., N$ exists, where the data points $\hat{x}_i \in \mathcal{X}_i$. Thus the sets $\mathcal{P}$ and $\mathcal{X}_i$ encode the uncertainty in the model parameters and the uncertainty in the data due to experimental error, respectively. We assume that these sets are *semi-algebraic*, i.e., that they can be described by a finite set of polynomial inequalities. For example, if $\hat{x}_1^{(i)} \in \left[\underline{\hat{x}_1^{(i)}}, \overline{\hat{x}_1^{(i)}} \right]$ for $i = 1, ..., n$, where $\hat{x}_1^{(i)}$ refers to the ith element of the experimental data taken at time t_1, then we obtain the n-dimensional hypercube:

$$\mathcal{X}_1 = \left[\hat{x}_i \in \mathcal{R}^n | \left(\hat{x}_1^{(i)} - \underline{\hat{x}_1^{(i)}} \right) \left(\hat{x}_1^{(i)} - \overline{\hat{x}_1^{(i)}} \right) \leq 0, \quad i = 1, ..., n \right] \qquad (6.13)$$

To invalidate this model, using this set of data, we need to show that no choice of model parameters from the set $\mathcal{P}$ will allow the model to match any data point in the set $\mathcal{X}_i$, i.e. that the set of measured experimental observations is incompatible with the "set" of models defined by $\mathcal{P}$. Note that in order to invalidate a model, one data point at $t = \mathcal{L}$ where $\mathcal{L} \in \{2, ..., N\}$, together with the initial time point t_1, is sufficient (usually the point with the largest residual between the nominal model and the data is selected).

The above problem can be solved using SOS programming via a method similar in concept to that of constructing a Lyapunov function to establish equilibrium stability. Lyapunov functions ensure the stability property of a system by guaranteeing that the state trajectories do not escape their sub-level sets. In [40], the related concept of barrier certificates is introduced. These are functions of state, parameter and time, whose existence proves that the candidate model is invalid given a parameter set and experimental data, by ensuring that the model behaviour does not intersect the set of experimental data. Consider a system of the form given in Eq. (6.12), and assume that $x \in \mathcal{X} \in \mathcal{R}^n$. Given this information, if it can be shown that for all possible system parameters $p \in \mathcal{P}$ the model cannot produce a trajectory $x(t)$ such that $x(t_1) \in \mathcal{X}_1, x(t_{\mathcal{L}}) \in \mathcal{X}_{\mathcal{L}}$ and $x(t) \in \mathcal{X}$ for all $t \in [t_1, t_{\mathcal{L}}]$, then the model and parameter set are invalidated by $\mathcal{X}_1, \mathcal{X}_{\mathcal{L}}, \mathcal{X}$. This idea leads to the following result, [40]:

Given the candidate model (6.12) and the sets $\mathcal{X}_1, \mathcal{X}_{\mathcal{L}}, \mathcal{X}, \mathcal{P}$, suppose there exists a real valued function $B(x, p, t)$ that is differentiable with respect to x and t such that

$$B(x_{\mathcal{L}}, p, t_{\mathcal{L}}) - B(x_1, p, t_1) > 0, \quad \forall(x_{\mathcal{L}}, x_1, p) \in \mathcal{X}_{\mathcal{L}} \times \mathcal{X}_1 \times \mathcal{P},$$

$$\frac{\delta B(x, p, t)}{\delta x} f(x, p) + \frac{\delta B(x, p, t)}{\delta t} \leq 0, \quad \forall(x, p, t) \in \mathcal{X} \times \mathcal{P} \times [t_1, t_{\mathcal{L}}].$$

Then the model is invalidated by $\mathcal{X}_1, \mathcal{X}_{\mathcal{L}}, \mathcal{X}$ and the function $B(x, p, t)$ is called a barrier certificate.

A key advantage of SOS programming is that these barrier certificates can be constructed algorithmically using Semidefinite Programming and SOS-TOOLS software. Using this approach, it was shown in [11] how a barrier certificate could be constructed for a simple generic biochemical network model, hence invalidating the model over a certain range of its parameters for a given set of time-course data, while in [43] it was shown how the same approach could be used to test a model of G-protein signalling in yeast. In [44] SOS tools were employed for the design of input experiments which maximise the difference between the outputs of two alternative models of bacterial chemotaxis. This approach can be used to design experiments to produce data that are most likely to invalidate incorrect model structures.

The main advantages of the SOS approach are that it can be applied to nonlinear models and that it is simulation-free, i.e. the results are analytical and thus provide deterministic guarantees. This is in contrast to simulation-based approaches which, for example, can never "prove" that a model with a given set of uncertain parameters will not enter a defined region of state-space (although of course in practice one can obtain answers to such questions with arbitrarily high statistical confidence if one is prepared to run enough simulations — see below). The main limitation of SOS techniques, aside from certain restrictions they place on the form of the model equations, is due to the computational limitations of the semidefinite programming software, which currently prohibits their application to high-order models.

6.2.6   Monte Carlo simulation

Monte Carlo simulation has for many years been the method of choice in the engineering industry for examining the effects of uncertainty on complex simulation models. The method is extremely simple, and relies on repeated simulation of the system over a random sampling of points in the model's parameter space. The sampling of the system's parameter space is usually carried out according to a particular probability distribution; for example, if there are reasons to believe that it is more likely for the system's actual parameter values to be near the nominal model values than to be near their uncertainty bounds, then a normal distribution may be used, whereas if no such information is available a uniform distribution may be chosen. For a given number of samples of a system's parameter space, statistical results can be derived which may be used to evaluate the effects of uncertainty on the system's behaviour. For the purposes of robustness analysis, these results pro-vide probabilistic confidence levels that the extremal behaviour found among the Monte Carlo simulations is within some distance of the true "worst-case" behaviour of the system.

The numbers of Monte Carlo simulations required to achieve various lev-els of estimation uncertainty with different confidence levels were calculated using the Chebyshev inequality and central limit theorem in [45] and are re-produced here in Table 6.1. Alternatively, if we use the well-known Chernoff bound, [46, 47], to estimate the number of simulations required, the numbers are as shown in Table 6.2. Note that in both cases it is clear that the number of samples required to produce a given set of statistical results is *indepen-dent* of the number of uncertain parameters in the model, and this, together with the absence of any requirements on the form of the model, represents the main advantage of Monte Carlo simulation for robustness analysis. The key disadvantage of the approach, however, is also readily apparent from the tables, namely, the exponential growth in the number of simulations with re-spect to the statistical confidence and accuracy levels required — typically at least 1000 simulations would be required in engineering applications before the statistical performance guarantees would be considered reliable. Although the statistical nature of the results generated using Monte Carlo simulation can sometimes hinder the comparison of the robustness properties of different models, one very useful capability of this approach is that it allows the char-acterisation of the size and shape of robust or non-robust regions of parameter

TABLE 6.1

Numbers of simulations for various confidence and accuracy levels
(derived using the Chebyshev inequality and central limit theorem, [45])

Percent of estimation uncertainty	20%	15%	10%	5%	1%
Uncertainty probability range					
0.750 → 0.954	25	45	100	400	10,000
0.890 → 0.997	57	100	225	900	22,500
0.940 → 0.999	100	178	400	1,600	40,000

TABLE 6.2

Numbers of Monte Carlo simulations required for
various confidence and accuracy levels (derived using
the Chernoff bound, [46])

% Confidence	Accuracy level ϵ	No. of simulations
99%	0.05	1,060
99.9%	0.01	27,081
99.9%	0.005	108,070

space. This is often an important issue in robustness analysis, since it is clear
that a model which fails a robustness test due to a single (perhaps biologically
unrealistic) parameter combination should not be considered equivalent to a
model which contains a large region of points which fail the same test. For
example, in [37], Monte Carlo simulation was used to establish that the loss of
oscillatory behaviour of a biochemical network model was not due to a single
point but to a significant region in its parameter space. In [48], the robustness
of models of the direct signal transduction pathway of receptor-induced apoptosis was evaluated via Monte Carlo simulation. By analysing the topology of
robust regions of parameter space, the robustness of the bistable threshold between cell reproduction and death could be evaluated in order to discriminate
between competing models of the network.

6.3 New robustness analysis tools for biological systems

The growth in interest in the notion of robustness in systems biology research
over the last decade has been remarkable and must represent one of the most
striking examples of the wholesale transfer of an idea from the field of engineering to the life sciences. Along with this interest in biological robustness
per se has come the recognition that many of the tools and methods that have

been developed within engineering to analyse the robustness of complex systems can be usefully employed by systems biologists in their efforts to develop and validate computational models. In a pleasing example of interdisciplinary feedback, this interest has recently spurred the development of several new techniques which are specifically oriented towards the analysis of biological systems.

In [49], for example, a computational approach was developed to investigate generic topological properties leading to robustness and fragility in large-scale biomolecular networks. This study found that networks with a larger number of positive feedback loops and a smaller number of negative feedback loops are likely to be more robust against perturbations. Moreover, the nodes of a robust network subject to perturbations are mostly involved with a smaller number of feedback loops compared with the other nodes not usually subject to perturbations. This topological characteristic could eventually make the robust network fragile against unexpected mutations at the nodes which had not previously been exposed to perturbations. In [50, 51], novel analytical approaches were developed for estimating the size and shape of robust regions in parameter space, which could provide useful complements or alternatives to traditional Monte Carlo analysis.

An evolutionary perspective on the generation of robust network topologies is provided in [14], where several hundred different topologies for a simple biochemical model of circadian oscillations were investigated *in silico*. This study found that the distribution of robustness among different network topologies was highly skewed, with most showing low robustness, with a very few topologies (involving the regulatory interlocking of several oscillating gene products) being highly robust. To address the question of how robust network topologies could have evolved, a topology graph was defined, each of whose nodes corresponds to one circuit topology that shows circadian oscillations. Two nodes in this graph are connected if they differ by only one regulatory interaction within the circuit. For the circadian oscillator under consideration, it could be shown that most topologies are connected in this graph, thus facilitating evolutionary transitions from low to high robustness. Interestingly, other studies of the evolution of robustness in biological macromolecules have generated similar results, suggesting that the same principles may govern the evolution of robustness on different levels of biological organisation.

A series of recent papers has introduced the notion of "flexibility" as an important counterpoint to robustness, particularly in the context of circadian clocks, [52, 53]. Flexibility measures how readily the rhythmic profiles of all the molecular clock components can be altered by modifying the biochemical parameters or environmental inputs of the clock circuit. Robustness, on the other hand, describes how well a biological function, such as the phase of a particular clock component, is maintained under varying conditions. As noted in [52, 53], the relationship between these two high-level properties can be a rather complex one, depending on the particular properties of the system of interest. This is because, although flexibility might be assumed to imply decreased robustness by increasing sensitivity to perturbations, in certain cases it can also yield greater robustness by enhancing the ability of the network to tune key environmental responses. This somewhat paradoxical result was nicely illustrated through the analysis of a model of the fungal circadian clock, which is based on the core FRQ-WC oscillator that incorporates both negative frq and positive wc-1 loops, as well as part of the light-signalling pathway. By introducing a simple measure of the flexibility of the network, based on quantifying how outputs of the entrained clock vary under parameter perturbations achievable by evolutionary processes, the authors demonstrate that the inclusion of the positive wc-1 feedback loop yields a more flexible clock. This increased flexibility is shown to be primarily characterised by a greater flexibility in entrained phase, leading to *enhanced* robustness against photoperiod fluctuations.

Another fundamental topic in systems biology is the effect of intrinsic stochastic noise on the stability of biological network models. Promising initial adaptations of traditional control engineering analysis techniques to address this issue were recently reported in [54, 55], and there is clearly tremendous scope for extending these results to deal with related robustness analysis problems.

The outlook for future research in this area is very positive, as the range of biological systems to which the approach to model validation outlined in this chapter is applied will no doubt continue to grow. This process will necessitate the development of new robustness analysis tools which can handle models which do not fall into the traditional category of differential equation-based systems, e.g. Boolean network models, Bayesian networks, hybrid dynamical systems, etc. As usual, progress is likely to be most rapid on the interface between traditionally separate domains of expertise, e.g. statistics and dynamical systems, [56], or evolutionary theory and control theory, [57, 58].

6.4   Case Study IX: Validating models of cAMP oscillations in aggregating *Dictyostelium* cells

A series of recent papers has used robustness analysis to interrogate and extend a model, originally proposed in [59] and shown in Fig. 2.23, of the biochemical network underlying stable oscillations in cAMP in aggregating *Dictyostelium* cells.

The dynamics of this network model, which is described in detail in Case Study II, were shown in [59] to closely match experimental data for the period, relative amplitudes and phase relationships of the oscillations in the concentrations of the molecular species involved in the network. Based on ad hoc simulations, the model was also claimed to be robust to very large changes in the values of its kinetic parameters, and this robustness was cited as a key advantage of the model over previously published models in the literature. However, a formal analysis of the robustness of the model to simultaneous variations in the values of its kinetic constants, using the structured singular value μ and global nonlinear optimisation, revealed extremely poor robustness characteristics, [37], as shown in Fig. 6.3. This rather surprising result merited further investigation in a number of follow-up studies, since the experimental justification for the proposed network structure appeared sound.

The first of these studies, [60], used Monte Carlo simulation to evaluate the effects of intrinsic stochastic noise, as well as the effects of synchronisation between individual *Dictyostelium* cells, on the robustness of the resulting cAMP oscillations. Interestingly, the effect of intrinsic noise was to *enhance* the robustness of cAMP oscillations to variations between cells, while synchronisation of oscillations between cells via a shared pool of external cAMP also significantly improved the robustness of the system. Finally, two further studies suggested a significant role for other subnetworks involving calcium and IP3 in generating robust oscillations, [61, 62]. Using a combination of structural robustness analysis [61] and biophysical modelling [62], an extended model including these subnetworks (Fig. 6.4) was constructed which exhibited significantly higher robustness than the original model, as shown in Fig. 6.5. The results of these studies clearly illustrate the power of robustness analysis techniques to analyse, develop and refine computational models of biochemical networks.

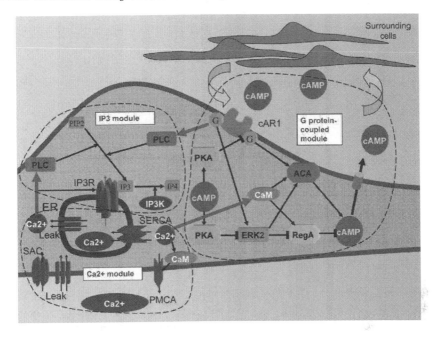

FIGURE 6.4: An extended model of the *Dictyostelium* cAMP oscillatory network incorporating coupled sub-networks involving Ca2+ and IP3,[62] — reproduced by permission of the Royal Society of Chemistry.

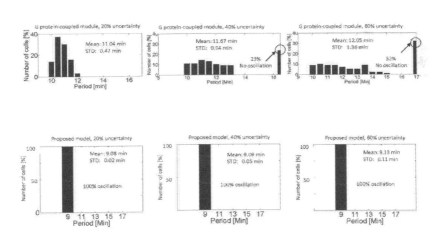

FIGURE 6.5: A comparison of the robustness of the original and extended model to variations in four kinetic parameters common to both models. Analysis was conducted using Monte Carlo simulations with three different levels of parametric uncertainty.

6.5   Case Study X: Validating models of the p53-Mdm2 System

Several recent studies have attempted to develop computational models of the complex dynamics of the p53-Mdm2 system. In [63], the authors developed a model in which ATM, a protein that senses DNA damage, activates p53 by phosphorylation. Activated p53 is modelled as having a decreased degradation rate and an enhanced transactivation of Mdm2. The model includes two explicit time delays, the first representing the processes (primarily, elongation and splicing) underlying the transcriptional production of mature, nuclear Mdm2 mRNA, and the second representing Mdm2 transport to the cytosol, translation to protein and transport of Mdm2 protein into the nucleus. As part of the model development process, the authors examined a large number of variations in their model to evaluate its robustness. For example, they explored other kinetics for ATM activation of p53 and Mdm2 ubiquitination of p53 and considered the effects of adding both Mdm2-dependent and Mdm2-independent ubiquitination of active p53. In all cases, the model was shown to be robust to such changes, and the conclusions arising from its analysis did not change. An investigation of the effects of varying different model parameters was carried out using bifurcation analysis, and this analysis produced new predictions regarding the source of robustness in the oscillatory dynamics. For example, with activated ATM-stimulated Mdm2 degradation, sustained oscillations occurred in the model if the total time delay was more than a 16 minute threshold. When the activated ATM-dependent degradation of Mdm2 was removed, however, while keeping the rest of the model parameters at their nominal values, then there were no sustained oscillations regardless of how high the time delay and the DNA damage was. Thus, the mechanism of activated ATM-dependent degradation of Mdm2 appears to be a key factor in ensuring oscillatory robustness in this system.

Another recent study of the p53 system considered six different mathematical models of the p53Mdm2 system, [64]. All of the models include the negative feedback loop in which p53, denoted by x, transcriptionally activates Mdm2, denoted by y, and active Mdm2 increases the degradation rate of p53. Three of the models were delay oscillators: Model I includes an Mdm2 precursor representing, for example, Mdm2 mRNA, and the action of y on x is described by first-order kinetics in both x and y. In model IV, the action of y on x is nonlinear and is described by a saturating Michaelis–Menten function. In model III, the Mdm2 precursor is replaced by a stiff delay term, which makes the production rate of Mdm2 depend directly on the concentration of p53 at an earlier time. Note that the model of [63] described above combines features of models III and IV. In addition to the three delay oscillators, the authors also considered two relaxation oscillators (II and V) in which the negative feedback loop is supplemented by a positive feedback loop on p53.

This positive feedback loop might represent in a simplified manner the action of additional p53 system components, which have a total upregulating effect on p53. These models include both linear positive regulation (model V) and nonlinear regulation based on a saturating function (model II). Models I–V, although differing in detail, all rely on a single negative feedback loop. The last model (VI) considered in the study proposes a novel checkpoint mechanism, which uses two negative feedback loops, one direct feedback and one longer loop that impinges on an upstream regulator of p53. In this model, a protein downstream of p53 inhibits a signaling protein that is upstream of p53.

In order to discriminate between these six different models of the p53 system, all six models were numerically solved for a wide range of parameter values and their robustness was evaluated. Models I–III were shown to be incapable of robustly producing stable undamped oscillations, while, in contrast, models IV–VI could generate sustained or weakly damped oscillations over a broad range of parameter values. Interestingly, most of the parameters shared by these three models showed very similar best-fit values, indicating that these models may provide estimates of the effective biochemical parameters such as production rates and degradation times of p53 and Mdm2. When low-frequency multiplicative noise was added to the protein production terms in the model to take account of stochasticity in protein production rates, all models showed qualitatively similar dynamics to those found in experiments, including occasional loss of a peak. However, only model VI was able to reproduce the authors' experimental observations that p53 and Mdm2 peak amplitudes had only a weak correlation (all other models had a strong coupling in the variations of the peaks of these two proteins).

Finally, a recent study of the robustness of the p53 protein-interaction network, [65], shows that the idea of robustness analysis can also be usefully applied at the topological network level. By subjecting the model to both random and directed perturbations representing stochastic gene knockouts from mutation during tumourigenesis, the p53 cell cycle and apoptosis control network could be shown to be inherently robust to random knockouts of its genes. Importantly, this robustness against mutational perturbation was seen to be provided by the structure of the network itself. This robustness against mutations, however, also implies a certain fragility, as the reliance on highly connected nodes makes it vulnerable to the loss of its hubs. Evolution has produced organisms that exploit this very weakness in order to disrupt the cell cycle and apoptosis system for their own ends: tumour inducing viruses (TIVs) target specific proteins to disrupt the p53 network, and this study identified these same proteins as the network hubs. Although TIVs had previously been likened to "biological hackers," this study showed why the TIV attack is so effective: TIVs target a specific vulnerability of the network that can be explained by analysing the robustness of the network architecture.

References

[1] Csete ME and Doyle JC. Reverse engineering of biological complexity. *Science*, 295:1664–1669, 2002.

[2] Carlson JM and Doyle JC. Complexity and robustness. *PNAS*, 99 (Suppl 1):2538–2545, 2002.

[3] Kurata H, El-Samad H, Iwasaki R, Ohtake H, Doyle JC, Grigorova I, Gross CA, and Khammash M. Module-based analysis of robustness tradeoffs in the heat shock response system. *PLoS Computational Biology*, 2(7):e59, DOI:10.1371/journal.pcbi.0020059, 2006.

[4] Kitano H. Cancer robustness: tumour tactics. *Nature*, 426:125 (2003).

[5] Kitano H. Biological robustness. *Nature Reviews Genetics*, 5:826–837, 2004.

[6] Kitano H and Oda K. Robustness trade-offs and host-microbial symbiosis in the immune system. *Molecular Systems Biology*, 2:2006.0022, 2006.

[7] Kitano H. Towards a theory of biological robustness. *Molecular Systems Biology*, 137:1–7, 2007.

[8] Kitano H. A robustness-based approach to system-oriented drug design. *Nature Reviews Drug Discovery*, 6:202–210, 2007.

[9] Wagner A. *Robustness and Evolvability in Living Systems*. Princeton: Princeton University Press, 2007.

[10] Morohashi M, Winnz AE, Borisuk MT, Bolouri H, Doyle JC and Kitano H. Robustness as a measure of plausibility in models of biochemical networks. *Journal of Theoretical Biology*, 216, 19-30, 2002.

[11] Anderson J and Papachristodoulou A. On validation and invalidation of biological models. *BMC Bioinformatics*, 10:132, 2009.

[12] Locke JCW, Southern MM, Kozma-Bognar L, Hibberd V, Brown PE, Turner MS, and Millar AJ. Extension of a genetic network model by iterative experimentation and mathematical analysis. *Molecular Systems Biology*, 1:2005.0013, 2005.

[13] Ueda HR, Hagiwara M, and Kitano H. Robust oscillations within the interlocked feedback model of Drosophila circadian rhythms. *Journal of Theoretical Biology*, 210(4):401–406, 2001.

[14] Wagner A. Circuit topology and the evolution of robustness in two-gene circadian oscillators. *PNAS*, 102(33):11775–11780, 2005.

[15] Akman OE, Rand DA, Brown PE, and Millar AJ. Robustness from flexibility in the fungal circadian clock. *BMC Systems Biology*, 4:88, DOI:10.1186/1752-0509-4-88, 2010.

[16] Thalhauser CJ and Komarova NL. Specificity and robustness of the mammalian MAPK-IEG Network. *Biophysical Journal*, 96:3471-3482, 2009.

[17] Yi T-M, Huang Y, Simon MI, and Doyle JC. Robust perfect adaptation in bacterial chemotaxis through integral feedback control. *PNAS*, 97(9):4649–4653, 2000.

[18] Ma L and Iglesias PA. Quantifying robustness of biochemical network models. *BMC Bioinformatics*, 3:38, 1–13, 2002.

[19] Battogtokh D and Tyson JJ. Bifurcation analysis of a model of the budding yeast cell cycle. *Chaos*, 14:653–661, 2004.

[20] Ghaemi R, Sun J, Iglesias PA, and Del Vecchio D. A method for determining the robustness of bio-molecular oscillator models. *BMC Systems Biology*, 3:95, 2009.

[21] Saltelli A et al. *Global Sensitivity Analysis: The Primer*. Chichester: Wiley-Interscience, 2008.

[22] Stelling J, Gilles ED, and Doyle FJ. Robustness properties of circadian clock architectures. *PNAS*, 101(36):13210–13215, 2004.

[23] Mahdavi A, Davey RE, Bhola P, Yin T, and Zandstra PW. Sensitivity analysis of intracellular signaling pathway kinetics predicts targets for stem cell fate control. *PLoS Computational Biology*, 3(7):e130. DOi:10.1371/journal.pcbi.0030130, 2007.

[24] Hafner M, Koeppl H, Hasler M, and Wagner A. Glocal robustness analysis and model discrimination for circadian oscillators. *PLoS Computational Biology*, 5(10):e1000534, DOI:10.1371/journal.pcbi.1000534, 2009.

[25] Doyle JC. Analysis of feedback systems with structured uncertainty. *IEE Proceedings on Control Theory and Applications, Part D*, 129(6): 242–250, 1982.

[26] Zhou K and Doyle, JC. *Essentials of Robust Control*. Upper Saddle River, N.J.: Prentice Hall, 1998.

[27] Skogestad S and Postlethwaite I. *Multivariable Feedback Control*. Chichester: John Wiley, 2nd edition, 2005.

[28] Ferreres, G. *A Practical Approach to Robustness Analysis with Aeronautical Applications*. New York: Kluwer Academic, 1999.

[29] Kim J, Bates DG, and Postlethwaite I. A geometrical formulation of the μ-lower bound problem. *IET Control Theory and Applications*, 3(4):465–472, 2009.

[30] Trane C and Jacobsen EW. Unraveling feedback structures in gene regulatory networks with application to the mammalian circadian clock. In *Proceedings of the 7th International Conference on Systems Biology (ICSB)*, Yokohama, Japan, 2006.

[31] Jacobsen EW and Trane C. Structural robustness of biochemical networks. In *Control Theory and Systems Biology*, B. Ingalls and P. Iglesias (Eds.), Boston: MIT Press, 2009.

[32] Jacobsen EW and Trane C. Using dynamic perturbations to identify fragilities in biochemical reaction networks. *International Journal of Robust and Nonlinear Control*, 20(9):1027–1046, 2010.

[33] Jacobsen EW and Cedersund G. Structural robustness of biochemical network models – with application to the oscillatory metabolism of activated neutrophils. *IET Systems Biology*, 2(1):39–47, 2008.

[34] Davis, L. *Handbook of Genetic Algorithms*. NewYork: Van Nostrand Reinhold, 1991.

[35] Jones DR, Perttunen CD, and Stuckman BE. Lipschitzian optimization without the Lipschitz constant. *Journal of Optimization Theory and Application*, 79:157–181, 1993.

[36] Menon PP, Postlethwaite I, Bennani S, Marcos A, and Bates DG. Robustness analysis of a reusable launch vehicle flight control law. *Control Engineering Practice*, 17:751–765, 2009.

[37] Kim J, Bates DG, Postlethwaite I, Ma L, and Iglesias P. Robustness analysis of biochemical network models. *IET Systems Biology*, 152(3):96–104, 2006.

[38] Miller JH. Active nonlinear tests (ANTs) of complex simulation models. *Management Science*, 44(6):820–830, 1998.

[39] Das A, Gao Z, Menon PP, Hardman JG, and Bates DG. A systems engineering approach to validation of a pulmonary physiology simulator for clinical applications. *Journal of the Royal Society Interface*, 8(54):44–55, doi:10.1098/rsif.2010.0224, 2011.

[40] Prajna S. Barrier certificates for nonlinear model validation. *Automatica*, 42(2):117–126, 2006.

[41] El-Samad H, Prajna S, Papachristodoulou A, Doyle JC, and Khammash M. Advanced methods and algorithms for biological network analysis. *Proceedings of the IEEE*, 94(4):832–853, 2006.

[42] Parillo P. Semidefinite programming relaxations for semialgebraic problems. *Mathematical Programming Series B*, 96(2):293–320, 2003.

[43] Yi TM, Fazel M, Liu X, Otitoju T, Goncalves J, Papachristodoulou A, Prajna S, and Doyle JC. Application of robust model validation using SOSTOOLS to the study of G-protein signalling in yeast. In *Proceedings of FOSBE*, 133–136, 2005.

[44] Melykuti B, August E, Papachristodoulou A, and El-Samad H. Discriminating between rival biochemical network models: three approaches to optimal experiment design. *BMC Systems Biology*, 4:38, 2010.

[45] Williams PS. A Monte Carlo dispersion analysis of the X-33 simulation software. In *Proceedings of the AIAA Conference on Guidance, Navigation and Control*, Paper No. 4067, 2001.

[46] Chernoff H. A measure of asymptotic efficiency for tests of a hypothesis based on the sum of observations. *Annals of Mathematical Statistics*, 23(4):493–507, 1952.

[47] Vidyasagar M. Statistical learning theory and randomised algorithms for control. *IEEE Control Systems Magazine*, 18(6):69–85, DOI: 10.1109/37.736014, 1998.

[48] Eissing T, Allgower F, and Bullinger E. Robustness properties of apoptosis models with respect to parameter variations and intrinsic noise. *IET Systems Biology*, 152(4):221–228, 2005.

[49] Kwon Y-K and Cho K-H. Quantitative analysis of robustness and fragility in biological networks based on feedback dynamics. *Bioinformatics*, 24(7):987–994, DOI:10.1093/bioinformatics/btn060, 2008.

[50] Chaves M, Sengupta A, and Sontag ED. Geometry and topology of parameter space: investigating measures of robustness in regulatory networks. *Journal of Mathematical Biology*, 59:315–358, 2009.

[51] Dayarian A, Chaves M, Sontag ED, and Sengupta A. Shape, size and robustness: feasible regions in the parameter space of biochemical networks. *PLoS Computational Biology*, 5:e10000256, 2009.

[52] Akman OE, Rand DA, Brown PE, and Millar AJ. Robustness from flexibility in the fungal circadian clock. *BMC Systems Biology*, 4:88, 2010.

[53] Edwards KD, Akman OE, Lumsden PJ, Thomson AW, Pokhilko A, Brown PE, Kozma-Bognar L, Nagy F, Rand DA, and Millar AJ. Quantitative analysis of regulatory flexibility under changing environmental conditions. *Molecular Systems Biology*, 6:424, 2010.

[54] Scott M, Hwa T, and Ingalls B. Deterministic characterization of stochastic genetic circuits. *PNAS*, 104(18):7402–7407, 2007.

[55] Kim J, Bates DG, and Postlethwaite I. Evaluation of stochastic effects on biomolecular networks using the generalised Nyquist stability criterion. *IEEE Transactions on Automatic Control*, 53(8):1937–1941, 2008.

[56] Kirk PD, Toni T, and Stumpf MP. Parameter inference for biochemical systems that undergo a Hopf bifurcation. *Biophysical Journal*, 95:540–549, ISSN:1542-0086(doi), 2008.

[57] Soyer OS and Pfeiffer T. Evolution under fluctuating environments explains observed robustness in metabolic networks. *PLoS Computational Biology*, 6(8):e1000907, DOI:10.1371/journal.pcbi.1000907, 2010.

[58] Salathe M and Soyer OS. Parasites lead to evolution of robustness against gene loss in host signaling networks. *Molecular Systems Biology*, 4:202, 2008.

[59] Laub MT and Loomis WF. A molecular network that produces spontaneous oscillations in excitable cells of *Dictyostelium*. *Molecular Biology of the Cell*, 9:3521–3532, 1998.

[60] Kim J, Heslop-Harrison P, Postlethwaite I, and Bates DG. Stochastic noise and synchronisation during *Dictyostelium* aggregation make cAMP oscillations robust. *PLoS Computational Biology*, 3(11):e218, doi:10.1371/journal.pcbi.0030218, 2007.

[61] Kim J-S, Valeyev NV, Postlethwaite I, Heslop-Harrison P, Cho K-W, and Bates DG. Analysis and extension of a biochemical network model using robust control theory. *International Journal of Robust and Nonlinear Control*, Special Issue on Robustness in Systems Biology: Methods and Applications, 20(9), DOI: 10.1002/rnc.1528, 2010.

[62] Valeyev NV, Kim J-S, Heslop-Harrison P, Postlethwaite I, Kotov N, and Bates DG. Computational modelling suggests dynamic interactions between Ca2+, IP3 and G protein-coupled modules are key to achieving robust *Dictyostelium* aggregation. *Molecular BioSystems*, 5:612-628, 2009.

[63] Ma L, Wagner J, Rice JJ, Hu W, Levine AJ, and Stolovitzky GA. A plausible model for the digital response of p53 to DNA damage. *PNAS*, 102(4):14266–14271, 2005.

[64] Geva-Zatorsky N, Rosenfeld N, Itzkovitz S, Milo R, Sigal A, Dekel E, Yarnitzky T, Liron Y, Polak P, Lahav G, and Alon U. Oscillations and variability in the p53 system. *Molecular Systems Biology*, DOI:10.1038/msb4100068, 2006.

[65] Dartnell L, Simeonidis E, Hubank M, Tsoka S, Bogle IDL, and Papageorgiou LG. Robustness of the p53 network and biological hackers. *FEBS Letters*, 579(14):3037–3042, 2005.

7

Reverse engineering biomolecular networks

7.1 Introduction

Fundamental breakthroughs in the field of biotechnology over the last decade, such as cDNA microarrays and oligonucleotide chips, [1, 2], have made high-throughput and quantitative experimental measurements of biological systems much easier and cheaper to make. The availability of such an overwhelming amount of data, however, poses a new challenge for modellers: how to reverse engineer biological systems at the molecular level using their measured responses to external perturbations (e.g. drugs, signalling molecules, pathogens) and changes in environmental conditions (e.g. change in the concentration of nutrients or in the temperature level). In this chapter, we provide an overview of some promising approaches, based on techniques from systems and control theory, for reverse engineering the topology of biomolecular interaction networks from this kind of experimental data. The approaches provide a useful complement to the many powerful statistical techniques for network inference that have appeared in the literature in recent years, [3].

7.2 Inferring network interactions using linear models

A standard approach to model the dynamics of biomolecular interaction networks is by means of a system of ordinary differential equations (ODEs) that describes the temporal evolution of the various compounds present in the system [4, 5]. Typically, the network is modelled as a system of rate equations in the form

$$\dot{x}_i(t) = f_i(x(t), p(t), u(t)), \qquad (7.1)$$

for $i = 1, \ldots, n$ with $x = (x_1, \ldots, x_n)^T \in \mathbb{R}^n$, where the state variables x_i denote the quantities of the different compounds (e.g. mRNA, proteins, metabolites) at time t, f_i is a function that describes the rate of change of the state variable x_i and its dependence on the other state variables, p is the parameter set and u is the vector of external perturbation signals.

The level of detail and the complexity of these kinetic models can be adjusted, through the choice of the rate functions f_i, by using more or less detailed kinetics, i.e. specific forms of f_i (linear or specific types of nonlinear functions). Moreover, it is possible to adopt a more or less simplified set of entities and reactions, e.g. choosing whether to take into account mRNA and protein degradation, or delays for transcription, translation and diffusion time [4]. When the order of the system increases, nonlinear ODE models quickly become intractable in terms of parametric analysis, numerical simulation and especially for identification purposes. Indeed, if the nonlinear functions f_i are allowed to take any form, determination of a unique solution to the inference problem becomes impossible even for quite small systems. Due to the above issues, although biomolecular networks are characterised by complex nonlinear dynamics, many network inference approaches are based on linear models or are limited to very specific types of nonlinear functions. This is a valid approach because, at least for small excursions of the relevant quantities from the equilibrium point, the dynamical evolution of almost all biological networks can be accurately described by means of linear systems, made up of ODEs in the continuous-time case, or difference equations in the discrete-time case (see [6, 7, 8, 9, 10, 11] and references therein).

Consider the continuous-time LTI model

$$\dot{x}(t) = Ax(t) + Bu(t), \tag{7.2}$$

where $x(t) = (x_1(t), \ldots, x_n(t))^T \in \mathbb{R}^n$, the state variables x_i, $i = 1, \ldots, n$, denote the quantities of the different compounds present in the system (e.g. mRNA concentrations for gene expression levels), $A \in \mathbb{R}^{n \times n}$ is the state transition matrix (the Jacobian of $f(x)$) and $B \in \mathbb{R}^{n \times 1}$ is a vector that determines the direct targets of external perturbations $u(t) \in \mathbb{R}$ (e.g. drugs, overexpression or downregulation of specific genes), which are typically induced during *in vitro* experiments. Note that the derivative (and therefore the evolution) of x_i at time t is directly influenced by the value $x_j(t)$ iff $A_{ij} \neq 0$. Moreover, the type (i.e. promoting or inhibiting) and extent of this influence can be associated with the sign and magnitude of the element A_{ij}, respectively. Thus, if we consider the state variables as quantities associated with the vertices of a directed graph, the matrix A can be considered as a compact numerical representation of the network topology. Since, in graph theory, two vertices are called *adjacent* when there is at least one edge connecting them, we can also denote A as the *weighted adjacency matrix* of the underlying network, where A_{ij} is the weight of the edge $j \to i$. Therefore, the topological reverse engineering problem can be recast as the problem of identifying the dynamical system (7.2). A possible criticism of this approach could be raised with respect to the use of a linear model, which is certainly inadequate to capture the complex nonlinear dynamics of certain molecular reactions. However, this criticism would be reasonable only if the aim was to identify an accurate model of large changes in the states of a biological system over time, and this is not the case here. If the goal is simply to describe the qualitative functional relationships between the states of the system when the system is subjected to perturbations, then a first-order linear approximation of the dynamics represents a valid choice of model. Indeed, a large number of approaches to network inference and model parameter estimation have recently appeared in the literature which are based on linear dynamical models, e.g. [6, 12, 9, 10, 13]. In addition to their conceptual simplicity, the popularity of such approaches arises in large part due to the existence of many well-established and computationally appealing techniques for the analysis and identification of this class of dynamical system.

Thus, the general problem of reverse engineering a biological interaction network from experimental data may be tackled via methods based on dynamical linear systems identification theory. The basic step of the inference process consists of estimating, from experimental measurements (either steady-state or time-series data), the weighted connectivity matrix A and the exogenous perturbation vector B of the *in silico* network model (7.2). Many different algorithms are available with which to solve this problem. The simplest approach is to use the classical least squares regression algorithm, which will be illustrated in Section 7.3.

7.2.1   Discrete-time vs continuous-time model

Since biological time-series data is always obtained from experiments at discrete sample points, when we identify the matrices $\hat{A}$ and $\hat{B}$ using this data we strictly speaking obtain not the estimates of A and B in Eq. (7.2), but rather those of the corresponding matrices of the discrete-time system obtained through the Zero-Order-Hold (ZOH) discretisation method ([14], p. 676) with sampling time T_s from system (7.2), that is

$$x(k+1) = A_d x(k) + B_d u(k), \tag{7.3}$$

where $x(k+1)$ is a shorthand notation for $x(kT_s + T_s)$, $x(k)$ for $x(kT_s)$, $u(k)$ for $u(kT_s)$, and

$$A_d = e^{AT_s}, \qquad B_d = \left(\int_0^{T_s} e^{A\tau} d\tau \right) B. \tag{7.4}$$

In general, the sparsity patterns of A_d and B_d differ from those of A and B. However, if the sampling time is suitably small, $(A)_{ij} = 0$ implies that $(A_d)_{ij}$ exhibits a very low value compared to the other elements on the same row and column, and the same applies for B_d and B. Therefore, in order to reconstruct the original sparsity pattern of the continuous-time system's matrices, one could set to zero the elements of the estimated matrices whose values are below a certain threshold.

In order to validate this approach, we will analyse more precisely the relationship between the dynamical matrices of the continuous-time and discrete-time systems. For the sake of simplicity, in what follows we will assume that

A has n distinct real negative eigenvalues, λ_i, $|\lambda_i| < |\lambda_{i+1}|$, $i = 1,\ldots,n$ and it is therefore possible to find a nonsingular matrix P such that* $A = PDP^{-1}$, with $D = \mathrm{diag}(\lambda_1,\ldots,\lambda_n)$. Then, the matrix A_d can be rewritten as ([15], p. 525)

$$A_d = I + AT_s + \frac{(AT_s)^2}{2!} + \frac{(AT_s)^3}{3!} + \cdots$$
$$= P\,\mathrm{diag}\left(e^{\lambda_1 T_s},\ldots,e^{\lambda_n T_s}\right)P^{-1}. \tag{7.5}$$

If the sampling time is properly chosen, such as to capture all the dynamics of the system, then $T_s \ll \tau_i := 1/|\lambda_i|$, $i = 1,\ldots,n$, which implies $|\lambda_i T_s| \ll 1$. Therefore the following approximation holds:

$$e^{\lambda_i T_s} = \sum_{k=0}^{\infty} \frac{(\lambda_i T_s)^k}{k!} \approx 1 + \lambda_i T_s.$$

From this approximation and Eq. (7.5), we obtain

$$A_d \approx I + AT_s.$$

As for the input matrix B, the following approximation holds:

$$B_d = A^{-1}\left(e^{AT_s} - I\right)B \approx A^{-1}(AT_s)B = BT_s$$

Note that the sparsity patterns of $I + AT_s$ and BT_s are identical to those of A and B, respectively. Only the diagonal entries of A can be significantly different from those of A_d. However, this is not an issue, because in all inference algorithms based on dynamical systems the optimisation parameters corresponding to the diagonal entries of A are always *a priori* assumed to be nonzero.

What can be concluded from the above calculations is that, in general, $(A)_{ij} = 0$ does not imply $(A_d)_{ij} = 0$; however, one can reasonably expect $(A_d)_{ij}$ to be much lower than the other elements on the i-th row and j-th column, provided that T_s is much smaller than the characteristic time constants of the system dynamics (the same applies for B and B_d). Such considerations can be readily verified by means of numerical tests, as illustrated in the following example.

Example 7.1

Consider a continuous-time linear dynamical system with five state variables and

$$A = \begin{pmatrix} -2.1906 & -0.7093 & 0 & 0 & 1.4131 \\ 0 & -2.2672 & -0.9740 & -0.0522 & 0 \\ 0 & -0.9740 & -4.0103 & 0 & 1.4374 \\ 0.4597 & -0.0522 & 0 & -1.8752 & 0 \\ 1.4131 & 0 & 0 & 0.1242 & -3.7822 \end{pmatrix}. \tag{7.6}$$

*The case of non-diagonalisable matrices is beyond the scope of the present treatment.

It is interesting to see what happens to the zero entries of A when the system

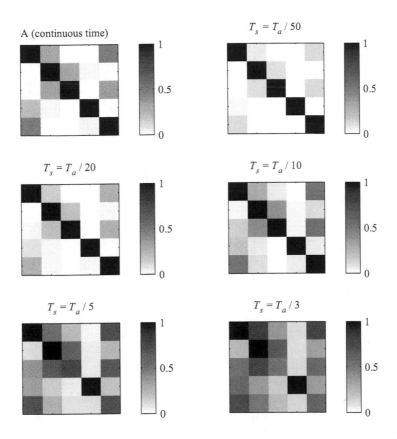

FIGURE 7.1: The zeros pattern of A and of its discretised versions (normalised matrices are shown) for different values of the sampling time. T_a is the settling time of the step response of the continuous-time system.

is discretised, for different values of the sampling time, using the ZOH transformation (7.4). The discrete-time versions of A are shown in Fig. 7.1: the pattern of the continuous-time A can be easily reconstructed when the sampling time T_s is small; indeed, the zero entries of A produce very small values in A_d. When T_s increases, the original pattern becomes hardly identifiable; moreover, all the values of A_d shrink towards zero. Note also that, when T_s is *too* small, the elements on the diagonal are much larger than the others. This is also problematic, because, as we will see later, small valued elements are more difficult to estimate when the data are noisy. We conclude that the sampling time plays a central role, no matter what inference technique will be used later on, and in general the optimal choice is a tradeoff between the need to capture the fastest dynamics of the system and the cost (both in terms of time and money) of having to make a larger number of measurements.   □

The algorithms presented in the next sections are based on the above arguments; indeed, each algorithm chooses at each step only the largest elements of the (normalised) estimated A_d and B_d matrices and is therefore expected to disregard the entries corresponding to zeros in the original matrix of the continuous-time model.

7.3   Least squares

Least Squares (LS) is by far the most widely used procedure for solving linear optimisation problems, especially in the field of identification. Assume that we are given h values of an independent vector variable, $x(k) \in \mathbb{R}^n$, and the corresponding measured values of a dependent scalar variable, $y(k) \in \mathbb{R}$, $k = 1, \dots, h$, obtained through the linear mapping

$$y(k) = \sum_{j=1}^{h} c_j \, x_j(k) + \nu(k) = c\,x(k) + \nu(k), \tag{7.7}$$

where the parameters c_j are unknown and ν represents the additive measurement white-noise term, that is with normal distribution, zero mean and σ^2 variance.

The LS method allows us to estimate the linear model that best describes the relationship between y and x, that is

$$\hat{y} = \sum_{j=1}^{h} \theta_j \, x_j = x^T \theta, \tag{7.8}$$

where $\hat{y}$ is the model estimate of y and $\theta \in \mathbb{R}^n$ is a vector of optimisation variables. In this context, the x_j are usually called *regressors* and the θ_j are called *regression coefficients*. If we define the error $e(k) := y(k) - \hat{y}(k)$, the quality of the approximation is measured in the least squares sense, i.e., a solution is optimal if it yields the minimal sum of squared errors $\sum_{k=1}^{h} e(k)^2$.

The problem can be conveniently reformulated in vector/matrix notation, by defining the following quantities:

$$X := \begin{pmatrix} x_1(1) & x_2(1) & \cdots & x_n(1) \\ x_1(2) & x_2(2) & \cdots & x_n(2) \\ \vdots & \vdots & & \vdots \\ x_1(h) & x_2(h) & \cdots & x_n(h) \end{pmatrix}, \; y := \begin{pmatrix} y(1) \\ y(2) \\ \vdots \\ y(h) \end{pmatrix}, \; \hat{y} := \begin{pmatrix} \hat{y}(1) \\ \hat{y}(2) \\ \vdots \\ \hat{y}(h) \end{pmatrix}, \; e := \begin{pmatrix} e(1) \\ e(2) \\ \vdots \\ e(h) \end{pmatrix}.$$

Now we can write the LS optimisation problem as

$$\min_{\theta} e^T e \tag{7.9a}$$

$$\text{s.t.} \quad e = y - \hat{y} = y - X\theta. \tag{7.9b}$$

Setting to zero the derivative of the loss function, $e^T e$, with respect to θ, one can easily derive the classical formula for the least squares estimate

$$\hat{\theta} = \left(X^T X\right)^{-1} X^T y. \tag{7.10}$$

The matrix $\left(X^T X\right)^{-1} X^T$ is called the (Moore–Penrose) *pseudo-inverse* of X and is often denoted by X^+. Note that, to compute X^+, it is necessary that $X^T X$ is invertible; this is possible if the n columns of X (the regression vectors) are linearly independent, which requires $h \geq n$, i.e., one should have at least as many measurements as regression coefficients. Note that, in theory, satisfaction of the latter inequality does not guarantee the invertibility of $X^T X$; however, this is always true in practice, because the presence of noise makes the probability of exact singularity equal to zero. On the other hand, a nonsingular $X^T X$ does not guarantee an accurate solution: when $X^T X$ is nearly singular the effects of noise and round-off errors on the estimated coefficients are very high, undermining the chances of recovering the true values.

If the real system is perfectly described by the model structure (7.7) and the data are not affected by noise ($\sigma^2 = 0$), then the optimal regression coefficients $\hat{\theta}_j$ coincide with the model parameters c_j. In practice, a linear model is often an approximation of the real system behaviour and the measurement noise is not negligible. Thus, it is interesting to investigate the relationship between the estimated regression coefficients and the actual coefficients. Some insight into the quality of the estimated model can be derived by inspecting the vector of *residuals*, defined as $y - \hat{y}$. A good model estimate should yield residuals that are close to white noise.

The accuracy of the estimated parameters can be described by their covariance matrix and it is possible to show that

$$\text{cov}(\hat{\theta}) = E\left\{\left(\hat{\theta} - E\{\hat{\theta}\}\right)\left(\hat{\theta} - E\{\hat{\theta}\}\right)^T\right\} = \sigma^2 \left(X^T X\right)^{-1}. \tag{7.11}$$

Since the diagonal entries of the matrix $\text{cov}(\hat{\theta})$ are the variances of the parameter estimates, Eq. (7.11) confirms the intuitive fact that the estimates are more accurate when the noise level is lower. Additionally, we can also conclude that, in general, the variance of the estimated parameters is smaller when the number of rows of X is higher: indeed, it is reasonable to assume that the absolute values of the entries of $X^T X$ increase linearly with h. Consequently, even in the presence of large amounts of noise, good estimates can still be obtained by increasing the number of measurements h.

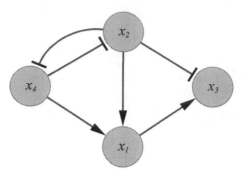

FIGURE 7.2: Toy network used in the examples. The shape of the arrow ending indicates the effect on the target node: $\triangle$ and $\top$ shapes are used for positive and negative effects, respectively (e.g. induction or repression of transcription of a gene).

Example 7.2

Let us consider the multivariable static linear relationship with additive noise

$$y = f(x, u) = Ax + Bu + v, \tag{7.12}$$

where

$$A = \begin{pmatrix} 0.7035 & 0.3191 & 0 & 0.0378 \\ 0 & 0.4936 & 0 & -0.0482 \\ 0.3227 & -0.4132 & 0.2450 & 0 \\ 0 & -0.3063 & 0 & 0.7898 \end{pmatrix}, \quad B = \begin{pmatrix} -1.2260 \\ 1.1211 \\ -1.1653 \\ 0.1055 \end{pmatrix} \tag{7.13}$$

and v is a vector of normally distributed random variables with zero mean and σ^2 variance. This is equivalent to four linear models in the form of Eq. (7.7), where the unknown parameters c_{ij}, $j = 1, \ldots, n+1$ of the i-th model are given by the i-th row of the matrix $[A\ B]$ and the independent vector variable is $z = [x^T\ u]^T$. System (7.12)-(7.13) can be associated with an interaction network of four nodes, whose topology is represented by the digraph in Fig. 7.2. Note that the matrix A describes the interactions between the nodes, whereas the B vector identifies the targets of the external perturbation, which in this case is assumed to directly affect all the nodes of the network. Note that we are not considering the transient response of the system; rather, we assume that Eq. (7.12) yields the next state of the network, starting from an initial state x and subject to a constant perturbation $u = 1$. Assume that the perturbation experiment has been repeated twenty times, starting from random initial values of x. Letting the measurements and regression matrices be $Y, Z \in \mathbb{R}^{20 \times 4}$, we want to identify the parameters of the model

$$\hat{Y} = Z\Theta, \tag{7.14}$$

from which we will get $\hat{\Theta}^T \in \mathbb{R}^{n \times (n+1)}$ as an estimate of $[A\,B]$. Denote by $\hat{A}$ the solution found by means of the LS formula (7.10). To evaluate such an estimate in terms of network inference we have to normalise each element, dividing it by the geometric mean of the norms of the row and column containing that element. Thus, we compute the normalised estimated adjacency matrix

$$\tilde{A}_{ij} = \frac{\hat{A}_{ij}}{\left(\|\hat{A}_{\star j}\| \cdot \|\hat{A}_{i\star}\| \right)^{1/2}} \qquad (7.15)$$

where $\hat{A}_{i\star}$ and $\hat{A}_{\star j}$ are the i-th row and j-th column of $\hat{A}$. When the noise is nonzero, all the elements of $\tilde{A}$ are usually nonzero as well. How can we translate this estimated matrix into an inferred network? The natural choice is to sort the list of edges in descending order according to the absolute value of their corresponding estimated parameters. Then, the elements at the top of the list will correspond to high-confidence predictions, i.e., edges with high probability of actually existing in the original network. This strategy is based on the idea that small perturbations of the experimental data should cause small variations in the coefficients, hence the zero entries of A should be identified by values that are close to zero.

In order to provide a statistically sound confirmation of this assumption, we can apply the LS-based identification procedure on a large number of experiments conducted on system (7.12)-(7.13). Repeating the same experiment many times allows us to compute a reliable average performance metric and to estimate the variability introduced by the random choice of x and by the additive measurement noise. Drawing the median absolute value of $\tilde{A}$ as a colormap allows us to effectively compare it with the normalised original adjacency matrix — see Fig. 7.3. Let us first consider the case $\sigma = 0.05$; focussing on the off-diagonal elements (the diagonal ones are always assumed to be different from zero), we note that below the diagonal the results are quite good, whereas there is some mismatch in the upper-right block of the matrix. This can be intuitively explained by the relatively small original values of the coefficients (1,4) and (2,4) (weights of the edges $4 \to 1$ and $4 \to 2$), which render their estimation more difficult. In general, we can conclude that, as it would be reasonable to expect, it is easier to infer the edges with larger weights.

We can now ask what happens if the noise increases so that $\sigma = 0.3$. From Fig. 7.3 it is evident that, while the *strong* edges are still inferred with good confidence, some new (wrong) low-confidence predictions appear. Eventually, if the noise is further increased (panel D), the estimated matrix becomes hardly useful in terms of network inference and we will obtain many wrong predictions. Finally, we can visualise the variability introduced by noise on each optimisation parameter: the box plots in Fig. 7.4 show that the variability is much higher when the noise is higher. Due to this high variability, the relative sorting of the parameters according to their absolute value is more likely to change between experiments. Thus, the probability of obtaining

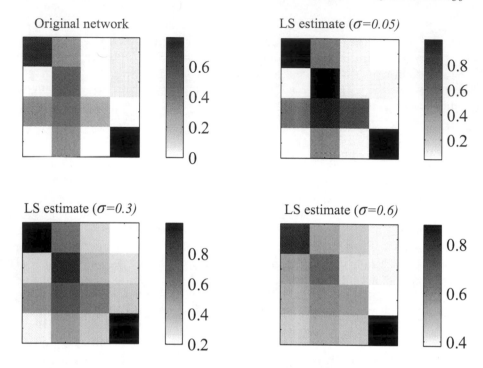

FIGURE 7.3: Median of the absolute values of the estimated $\hat{A}$ and of the normalised original adjacency matrix in Example 7.2. Median obtained over 100 experiments, with 20 z-y pairs for each experiment and different noise standard deviations.

wrong predictions when using the LS on a single experiment is fairly high in the latter case, whereas it is almost zero (at least for the first four predictions) when $\sigma = 0.05$. □

7.3.1   Least squares for dynamical systems

So far, we have applied the LS method to estimate the parameters of a static relationship between a vector of assigned independent variables, x, and a dependent variable, y, from noisy measurements. However, in many cases we cannot neglect the fact that the measured quantities are evolving in time, that is they are the state variables of an underlying dynamical system. Hence, we would like to have a method for network inference based on time-series data. In the following, we show how (and to what extent) it is possible to exploit LS to estimate a dynamical model of a biomolecular interaction network and its topology.

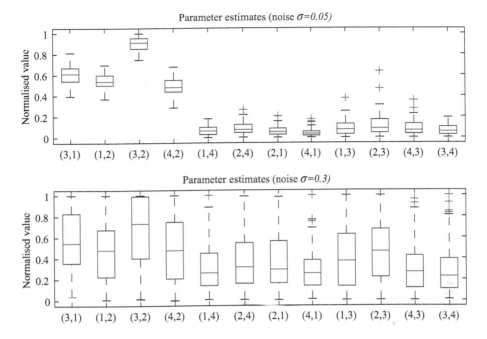

FIGURE 7.4: Distribution of the normalised values of the optimised parameters over 100 experiments for two different noise levels (Example 7.2). The first six box plots from the left are those corresponding to the actually existing edges of the original network; the second six are incorrect predictions (labels on the x-axis denote the row and column indexes in the $\tilde{A}$ matrix).

Assume the linearised discrete-time dynamical model of our network is

$$x(k+1) = A_d x(k) + B_d u(k) \tag{7.16}$$

and that $h+1$ experimental observations, $x(k) \in \mathbb{R}^n$, $k = 0, \ldots, h$, are available. Let

$$\hat{Y} := \begin{pmatrix} x_1(h) & x_2(h) & \ldots & x_n(h) \\ x_1(h-1) & x_2(h-1) & \ldots & x_n(h-1) \\ \vdots & \vdots & & \vdots \\ x_1(1) & x_2(1) & \ldots & x_n(1) \end{pmatrix}, \tag{7.17}$$

$$\hat{Z} := \begin{pmatrix} x_1(h-1) & x_2(h-1) & \ldots & x_n(h-1) & u(h-1) \\ x_1(h-2) & x_2(h-2) & \ldots & x_n(h-2) & u(h-2) \\ \vdots & \vdots & & \vdots & \vdots \\ x_1(0) & x_2(0) & \ldots & x_n(0) & u(0) \end{pmatrix}. \tag{7.18}$$

The identification model is then

$$\hat{Y} = \hat{Z}\Theta, \tag{7.19}$$

where $\Theta \in \mathbb{R}^{n \times (n+1)}$ is the optimisation matrix and the estimate of system (7.16) is given by $\hat{\Theta}^T = [\hat{A}_d \quad \hat{B}_d]$. Since Eq. (7.19) has the same structure as Eq. (7.14), we are naturally led to apply the LS formula (7.10) to solve it. Although, in principle, the solution is correct, some care must be taken due to the intrinsic differences between the two problems.

The first thing to notice is that the regressor matrix is not made up of independent variables, as in the static case: the columns of $\hat{Z}$ include the state vectors at the steps $0, 1, \ldots, h-1$, while the columns of $\hat{Y}$ are the same state vectors, but shifted one step ahead. A second point, which stems from the first, is that, in the LS formulation for dynamical system identification, the regressor variables are affected by noise, whereas in the static case they are deterministic. For this reason, Eq. (7.11) is no longer valid and we lack an estimate of the parameters' variance. A final consideration concerns the correlation between the regressor columns of $\hat{Z}$: examining Eq. (7.16) and looking at a typical step response of a dynamical system (see Fig. 7.5), we can clearly see that the value of the state vector at the k-th step is dependent on the value at the previous step. If the dynamics of the system are smooth and slow, then $x(k)$ can be approximated by a linear combination of its values at the previous step, $x(k-1), \ldots, x(0)$. This is quite unfortunate, because it means the columns of $Z^T Z$ are almost linearly dependent, which, as we have seen, renders the LS solution highly sensitive to noise and numerical round-off errors.

Example 7.3
In order to compare the effectiveness of the LS algorithm in the static and dynamical system cases, let us consider a dynamical system in the form of Eq. (7.16), with A and B given by the same matrices (7.13) used for the static system identification in Example 7.2. Figs. 7.6-7.7 show the results obtained by computing the LS solution to Eq. (7.19) for different noise levels: the performance is clearly worse compared to the analogous experiments conducted on the static system (7.12) (Fig. 7.3). The effect of noise on the median identified parameter values and on their variances is already significant at $\sigma = 0.05$; at $\sigma = 0.3$ the chance of recovering the true edges is almost equal to that obtained by a random guess.   ☐

A final comment is in order regarding the possibility of improving the identification results when using time-series measurements: for static systems identification, Eq. (7.11) suggests that, to obtain better estimates, one can increase the number of measurements. In the dynamical systems case, this could induce us to increase the number of measurements in the time-course experiments, by either reducing the sample time or by considering a longer

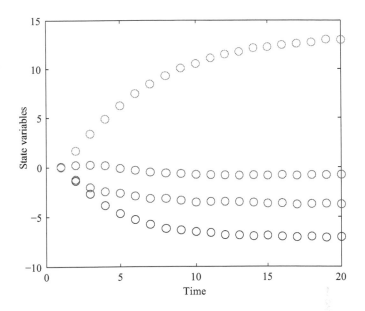

FIGURE 7.5: Step response of system (7.16), with the matrices A_d and B_d given by Eq. (7.13) and additive measurement noise $\sigma = 0.1$.

time interval. However, both these strategies are basically not useful: indeed, having $x(k)$ too close in time to $x(k-1)$ increases the approximate linear dependence between the regression vectors. On the other hand, taking additional measurements after the signals have reached the steady-state will again introduce new linearly dependent regression vectors (see Fig. 7.5, after step $k = 15$ the value of $x(k)$ is almost equal to $x(k-1)$). Hence, the only chance to improve the inference performance is by making many different experiments, possibly using different perturbation inputs which affect different nodes of the network.

7.3.2 Methods based on least squares regression

Many different algorithms for reverse engineering biomolecular networks based on the use of linear models and least squares regression have recently appeared in the literature, including NIR (Network Identification by multiple Regression, [6]), MNI (Microarray Network Identification, [16]) and TSNI (Time-Series Network Identification, [17, 18, 19]). The NIR algorithm has been developed for application with perturbation experiments on gene regulatory networks. The direct targets of the perturbation are assumed to be known and the method uses only the steady-state gene expression. Under the

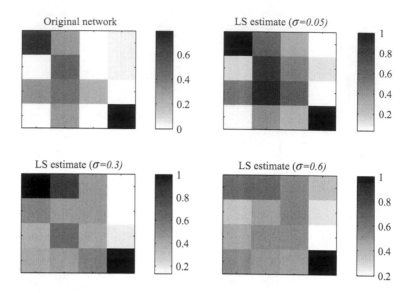

FIGURE 7.6: Median of the absolute values of the estimated $\hat{A}_d$ and of the normalised original adjacency matrix in Example 7.3. Median obtained over 100 experiments, with 20 z-y pairs for each experiment and different noise standard deviations.

steady-state assumption ($\dot{x}(t) = 0$ in Eq. (7.2)) the problem to be solved is

$$\sum_{j=1}^{n} a_{ij} x_j = -b_i u \tag{7.20}$$

The least squares formula is used to compute the network structure, that is the rows $a_{i,\star}$ of the connectivity matrix, from the gene expression profiles (x_j, $j = 1, \ldots, n$) following each perturbation experiment; the genes that are directly affected by the perturbation are expressed through a nonzero element in the B vector. NIR is based on a network sparsity assumption: only k (maximum number of incoming edges per gene) out of the n elements on each row are different from zero. For each possible combination of k out of n weights, the k coefficients for each gene are computed so as to minimise the interpolation error. The maximum number of incoming edges, k, can be varied by the user. An advantage of NIR is that k can be tuned so as to avoid underdetermined problems. Indeed, if one has N_e different (independent) perturbation experiments, the exact solution to the regression problem can be found for $k \leq N_e$, at least in the ideal case of zero noise.

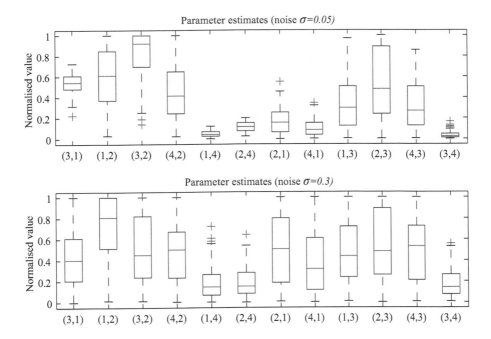

FIGURE 7.7: Distribution of the normalised values of the optimised parameters over 100 experiments for two different noise levels (Example 7.3). The first six box plots from the left are those corresponding to the actually existing edges of the original network, the second six are wrong predictions (labels on the x-axis denote the row and column indexes in the $\hat{A}$ matrix).

The MNI algorithm, similarly to NIR, uses steady-state data and is based on relation (7.20), but it does not require *a priori* knowledge of the specific target gene for each perturbation. The algorithm employs an iterative procedure: first, it predicts the targets of the treatment using a full network model; subsequently, it translates the predicted targets into constraints on the model structure and repeats the model identification to improve the reconstruction. The procedure is iterated until certain convergence criteria are met.

The TSNI algorithm uses time-series data, instead of steady-state values, of gene expression following a perturbation. It identifies the gene network (A), as well as the direct targets of the perturbations (B), by applying the LS to solve the linear equation (7.2). Note that, to solve Eq. (7.2), it is necessary to measure the derivative values, which are never available in practice. Also, numerical estimation of the derivative is not a suitable option, since it is well known to yield considerable amplification of the measurement noise. The solution implemented by TSNI consists of converting the system from continuous-time to discrete-time. The identification problem admits a unique globally optimal solution if $h \geq n + p$, where h is the number of data points, n is the number of state variables and p is the number of perturbations. To increase the number of data points, after using a cubic smoothing spline filter, a piecewise cubic spline interpolation is performed. Then a Principal Component Analysis (PCA) is applied to the data set in order to reduce its dimensionality and the problem is solved in the reduced dimension space. In order to compute the continuous-time system's matrices, A and B, from the corresponding discretised A_d and B_d, respectively, the following bilinear transformation is applied, [20]:

$$A = \frac{2A_d - I}{T_s A_d + I}$$
$$B = (A_d + I)AB_d$$

where $I \in \mathbb{R}^{n \times n}$ is the identity matrix and T_s the sampling interval.

Finally, the *Inferelator* technique, [21], also belongs to this category of algorithms. It uses regression and variable selection to infer regulatory influences for genes and/or gene clusters from mRNA and/or protein expression levels.

Two significant limitations which are common to almost all of the algorithms described above are their inability to (a) deal effectively with measurement noise in the experimental data, and (b) exploit prior qualitative or quantitative knowledge about parts of the network to be reconstructed. In the following sections, we describe promising new techniques, based on convex optimisation and extensions of the standard LS, which can address these issues.

7.4   Exploiting prior knowledge

A serious limitation of most methods based on LS regression is their inability to exploit any prior knowledge about the network topology in order to improve the inference performance. This is a major failing, since for any given network there is often a significant amount of information available in the biological literature and databases about certain aspects of its topology. For example, it is often possible to derive qualitative information about some part of a network from previously published experimental studies, e.g. "protein A inhibits the expression of gene B." When using standard regression techniques, a parameter can either be designated as a free optimisation variable or set to a constant value (i.e., a precise *quantitative* information). This makes it difficult to take into account qualitative *a priori* information, a problem which does not arise for statistical approaches such as Bayesian networks. Indeed, since for topological inference one often uses simplified models, the value of a certain model parameter does not necessarily correspond to a real experimentally measurable quantity. Moreover, at the biomolecular level accurate values of the system's parameters are seldom available or measureable and are often highly variable among different individuals of the same species. For these reasons, a much more suitable approach consists of exploiting the *a priori* available qualitative biological knowledge, by translating it into mathematical constraints on the optimisation variables; e.g. one can constrain a parameter to belong to the set of real positive (or negative) numbers, to be null or belong to an assigned interval.

In the following, we show how to recast the network inference as a convex optimisation problem using linear matrix inequalities (LMIs), [22, 11]. Similar approaches for identifying genetic regulatory networks using expression profiles from genetic perturbation experiments are described in [23] and [9, 10]. The distinctive feature of these approaches is that they easily enable the exploitation of any qualitative prior information which may be available from the biological domain, thus significantly increasing the inference performance. Furthermore, the wide availability of effective numerical solvers for convex optimisation problems renders this formalism very well-suited to deal with complex network inference tasks.

7.4.1 Network inference via LMI-based optimisation

Assuming that $h + 1$ experimental observations, $x(k) \in \mathbb{R}^n$, $k = 0, \ldots, h$, are available, our goal is to formulate the problem of estimating matrices A_d and B_d of system (7.16) in the framework of convex optimisation. In particular, we want to cast the problem as a set of LMIs.

Using the same notation as in Eq. (7.17)-(7.19), the identification problem can be transformed into that of minimising the norm of $\hat{Y} - \hat{Z}\Theta$, and thus we can state the following problem:
Given the sampled data set $x(k)$, $k = 0, \ldots, h$, and the associated matrices $\hat{Y}$, and $\hat{Z}$, find

$$\min_{\Theta} \varepsilon \qquad (7.21a)$$

$$\text{s.t.} \quad \left(\hat{Y} - \hat{Z}\Theta\right)^T \left(\hat{Y} - \hat{Z}\Theta\right) < \varepsilon I. \qquad (7.21b)$$

Note that condition (7.21b) is quadratic in the unknown matrix variable Θ. In order to obtain a linear optimisation problem, we convert it to the equivalent condition

$$\begin{pmatrix} -\varepsilon I & \left(\hat{Y} - \hat{Z}\Theta\right)^T \\ \left(\hat{Y} - \hat{Z}\Theta\right) & -I \end{pmatrix} < 0, \qquad (7.22)$$

by applying the properties of Schur complements (see [15], p. 123). The equivalence between Eq. (7.21b) and Eq. (7.22) is readily derived as follows.

Let $M \in \mathbb{R}^{n \times n}$ be a square symmetric matrix partitioned as

$$M = \begin{pmatrix} M_{11} & M_{12} \\ M_{12}^T & M_{22} \end{pmatrix}, \tag{7.23}$$

and assume that M_{22} is nonsingular. Defining the Schur complement of M_{22} as $\Delta := M_{11} - M_{12}M_{22}^{-1}M_{12}^T$, then the following statements are equivalent:

i) M is positive (negative) definite;

ii) M_{22} and Δ are both positive (negative) definite.

To see this, recall that M is positive (negative) definite iff

$$\forall x \in \mathbb{R}^n, \; x^T M x > 0 \quad (< 0),$$

and moreover it can be decomposed as ([24], p. 14)

$$\begin{aligned} M &= \begin{pmatrix} M_{11} & M_{12} \\ M_{12}^T & M_{22} \end{pmatrix} \\ &= \begin{pmatrix} I & M_{12}M_{22}^{-1} \\ 0 & I \end{pmatrix} \begin{pmatrix} \Delta & 0 \\ 0 & M_{22} \end{pmatrix} \begin{pmatrix} I & M_{12}M_{22}^{-1} \\ 0 & I \end{pmatrix}^T. \end{aligned}$$

The latter is a congruence transformation ([15], p. 568), which does not modify the sign definiteness of the transformed matrix; indeed, $\forall x \in \mathbb{R}^n$ and $\forall C, P \in \mathbb{R}^{n \times n}$

$$P \text{ positive (negative) definite} \Rightarrow x^T C^T P C x = z^T P z > 0 \quad (< 0).$$

Therefore M is positive (negative) definite iff M_{22} and Δ are both positive (negative) definite. Problem (7.21) with the inequality constraint in the form of Eq. (7.22) is a generalised eigenvalue problem ([25], p. 10), and can be easily solved using efficient numerical algorithms, such as those implemented in the MATLAB® LMI Toolbox [26].

A noteworthy advantage of the proposed convex optimisation formulation is that the approach can be straightforwardly extended to the case of multiple experimental data sets for the same biological network. In this case, there are several matrix pairs $(\hat{Y}^{(k)}, \hat{Z}^{(k)})$, one for each experiment: the problem can be formulated again as in Eq. (7.21), but using a number of constraints equal to the number of experiments, that is

$$\min_{\Theta} \sum_k \varepsilon_k$$

$$\text{s.t.} \quad \left(\hat{Y}^{(k)} - \hat{Z}^{(k)}\Theta \right)^T \left(\hat{Y}^{(k)} - \hat{Z}^{(k)}\Theta \right) < \varepsilon_k I, \quad k = 1, \ldots, N_e,$$

where N_e is the number of available experiments.

Except for the LMI formulation, the problem is identical to the one tackled by classical linear regression, that is finding the values of $n(n+1)$ parameters of a linear model that yield the best fitting of the observations in the least squares sense. Hence, if the number of observations, $n(h+1)$, is greater than or equal to the number of regression coefficients, that is $h \geq n$, the problem admits a unique globally optimal solution. In the other case, $h < n$, the interpolation problem is undetermined; thus, there exist infinitely many values of the optimisation variables that equivalently fit the experimental measurements. In the latter case, it is crucial to exploit clustering techniques to reduce the number of nodes and smoothing techniques to increase the number of samples, in order to satisfy the constraint $h \geq n$. Furthermore, adopting a bottom-up reconstruction approach (i.e. starting with a blank network and increasingly adding new edges) may help in overcoming the dimensionality problem: in this case, indeed, the number of edges incident to each node (and therefore the number of regression coefficients) is iteratively increased and can be limited to satisfy the above constraint.

As first noted in [22], the key advantage of the LMI formalism is that it makes it possible to take into account prior knowledge about the network topology by forcing some of the optimisation variables to be zero and other ones to be strictly positive (or negative), by introducing the additional inequality $A_{ij} > 0$ (< 0) to the set of LMIs. Similarly, we can impose a sign constraint on the i-th element of the input vector, b_i, if we *a priori* know the qualitative (i.e. promoting or repressing) effect of the perturbation on the i-th node. Also, an edge can be easily pruned from the network by setting to zero the corresponding entry in the matrix optimisation variable in the LMIs.

In the next subsections we present two iterative algorithms based on the convex optimisation approach described above.

The first algorithm prunes a fully connected network while the second algorithm implements the opposite approach: it starts with an empty network, then allows it to grow based on the mechanism of *preferential attachment*, [27]. According to this evolutionary mechanism, when a new node is added to the network it is more likely to establish a connection with a highly connected node (a hub) than with a loosely connected one.

7.4.2 MAX-PARSE: An algorithm for pruning a fully connected network according to maximum parsimony

The MAX-PARSE algorithm employs an iterative procedure: starting with a fully connected network, the edges are subsequently pruned according to a maximum parsimony criterion. The pruning algorithm terminates when the estimation error exceeds an assigned threshold. The following basic ideas underpin the pruning algorithm:

a) The optimal network model, among all those that yield an acceptably small error with respect to the experimental data, is the one with the minimum number of edges; this *maximum parsimony* criterion is based on the principle that nature optimises systems through evolution, aiming at the most efficient use of resources. Clearly this is a simplistic approach, because it does not take into account the fact that redundancy is implemented by many biological systems to achieve robustness.

b) Given the estimated (normalised) connectivity matrix at each iteration, the regression coefficients with low values correspond to non-adjacent (in the original network) nodes. Thus, these edges are the best candidates for pruning. This stems from the assumption that an indirect interaction typically results in a smaller weight in the rate equation of a certain species compared to the contributions of directly interacting species. This is also supported by the numerical experiments illustrated in Section 7.3.

Since the problem is formulated as a set of LMIs, the algorithm is also capable of directly exploiting information about some specific interactions that are *a priori* known, taking into account both the direction of the influence and its type (promoting or repressing). The reconstruction algorithm is structured in the following steps.

P1) A first system is identified by solving the optimisation problem (7.21) and adding all the known sign constraints.

P2) Let $\hat{A}^{(k)}$ be the matrix computed at the k-th step; in order to compare the elements of $\hat{A}^{(k)}$ we compute the normalised matrix $\tilde{A}^{(k)}$ according to Eq. (7.15).

P3) If the value $\tilde{A}_{ij}^{(k)}$ is below an assigned threshold, ε_p, the edge $j \to i$ is pruned and the corresponding regression coefficient is set to zero at the next iteration. This rule reflects the idea that an edge is a good candidate for elimination if its weight is low compared to the other edges arriving to and starting from the same node.

P4) A new LMI problem is cast, eliminating the optimisation variables chosen at the previous step, and a new solution is computed.

P5) The evolution of the identified system is compared with the experimental data: if the residual error exceeds a prefixed threshold, ε_{res}, then the algorithm stops; otherwise another iteration starts from point P2.

The algorithm requires tuning two optimisation parameters: the threshold ε_p, used in the pruning phase, which affects the number of coefficients eliminated at each step, and ε_{res}, defining the admissible estimation error, which determines the algorithm termination. The first parameter influences the connectivity of the final reconstructed network: the greater its value, the lower the final number of connections. The algorithm terminates when either it does not find any new edges to remove or the estimation error exceeds ε_{res}.

7.4.3 CORE-Net: A network growth algorithm using preferential attachment

Similarly to MAX-PARSE, the CORE-Net algorithm is based on the formulation of the network inference problem as a convex optimisation problem in the form of LMIs. However, the latter adopts a different heuristic to reconstruct the network topology: it uses an incremental reconstruction approach, starting with an empty network (no edges) and then iteratively adding new edges at each iteration.

The edges selection strategy implemented by CORE-Net is inspired by the experimental observation that the connectivity degree in metabolic [28], protein–protein interaction, [29] and gene regulatory networks, [30], as well as other genomic properties, [31], exhibits a power-law distribution. Roughly speaking, this means that only a small number of nodes (the *hubs*) are highly connected, whereas there are many loosely connected ones. A plausible hypothesis for the emergence of such a feature, as discussed in [27], is the *preferential attachment* (PA) mechanism, which states that during network growth and evolution the hubs have greater probability to establish new connections. In large networks, this evolution rule may generate particular degree distributions, such as the well-known power-law distribution that characterises scale-free networks.

Employing the PA mechanism within the reconstruction process, CORE-Net mimics the evolution of a biological network to improve the inference performance. Finally, it is worth noting that, while MAX-PARSE starts with a full adjacency matrix, CORE-Net progressively increases the number of regression coefficients. This mechanism tends to limit the final number of regression coefficients, in agreement with the maximum parsimony criterion used also in MAX-PARSE. This strategy is also effective in avoiding underdetermined estimation problems and in limiting the computational burden.

7.5 Dealing with measurement noise

The measurement error affecting the majority of biological experiments is substantial. The level of measurement noise is often difficult to determine, since it arises from different sources: 1) errors inherent in the measurement technique, 2) errors occurring at the time of sampling (with absolute and drift components) and 3) variability among different individuals of the same species. The effects of the resulting noise could, in principle, be limited by using more accurate measurement techniques and by data replication. These strategies, however, are in conflict with the applicability of high-throughput, fast and affordable measurement techniques, which is at the very base of the systems biology paradigm. Therefore, it is paramount to devise methods to address, in a principled manner, the network inference problem in the presence of substantial, but poorly defined, noise components. On the other hand, the development of such approaches can also provide valuable suggestions on how to optimise the experimental sampling strategies.

As noted in Section 7.3.1, the standard LS method is not capable of dealing effectively with noisy regressors. In the following we introduce two extensions, namely the Total Least Squares (TLS), [32, 33], and the Constrained Total Least Squares (CTLS), [34, 35], which have been developed to deal with this issue. Both of these algorithms are routinely used in advanced signal and image processing applications, and their usefulness in the context of systems biology is now also beginning to be appreciated.

7.5.1   Total least squares

Let us reconsider the formulation of the LS problem, by explicitly taking into account the additive noise terms in the measurements. The regression model for the i-th state variable becomes

$$Y_{\star i} + \Delta Y_{\star i} = (Z + \Delta Z) \cdot \Theta_{\star i} \,, \tag{7.24}$$

where $\Theta_{\star i} = \begin{pmatrix} a_{i1} & \cdots & a_{in} & b_i \end{pmatrix}^T$ is the vector of unknown parameters and

$$Y_{\star i} = \begin{pmatrix} x_i(h) \\ \vdots \\ x_i(1) \end{pmatrix}, \quad Z = \begin{pmatrix} x_1(h-1) & \cdots & x_n(h-1) & 1 \\ \vdots & \ddots & \vdots & \vdots \\ x_1(0) & \cdots & x_n(0) & 1 \end{pmatrix}$$

$$\Delta Y_{\star i} = \begin{pmatrix} \nu_i(h) \\ \vdots \\ \nu_i(1) \end{pmatrix}, \quad \Delta Z = \begin{pmatrix} \nu_1(h-1) & \cdots & \nu_n(h-1) & 0 \\ \vdots & \ddots & \vdots & \vdots \\ \nu_1(0) & \cdots & \nu_n(0) & 0 \end{pmatrix}$$

and $\nu_i(k)$ is the additive noise term on the i-th state variable at time step k. Although the exact values of the correction terms, ΔZ and $\Delta Y_{\star i}$, will not generally be known, the structure, i.e. how the noise appears in each element, can often be estimated.

First of all, let us write Eq. (7.24) in a more compact form, by defining

$$C^{(i)} := \begin{pmatrix} Z & Y_{\star i} \end{pmatrix},$$

$$\Delta C^{(i)} := \begin{pmatrix} \Delta Z & \Delta Y_{\star i} \end{pmatrix}.$$

Then Eq. (7.24) is rewritten as

$$\left(C^{(i)} + \Delta C^{(i)} \right) \begin{pmatrix} \Theta_{\star i} \\ -1 \end{pmatrix} = 0 \,. \tag{7.25}$$

The Total Least Squares (TLS) method computes the optimal regression parameters minimising the correction term ΔC. The TLS optimisation problem is posed as follows, [32]:

$$\min_{v,\Theta_{*i}} \|\Delta C^{(i)}\|_F^2$$

$$\text{s.t.} \quad \left(C^{(i)} + \Delta C^{(i)}\right) \begin{pmatrix} \Theta_{*i} \\ -1 \end{pmatrix} = 0, \tag{7.26}$$

where $\| \cdot \|_F$ denotes the Frobenius norm. When the smallest singular value of $C^{(i)}$ is not repeated, the solution of the TLS problem is

$$\Theta_{*i}^{\text{TLS}} = \left(Z^T Z - \lambda_i^2 I\right)^{-1} Z^T Y_{*i}, \tag{7.27}$$

where λ_i is the smallest singular value of $C^{(i)}$. Comparing Eq. (7.27) to the classical LS solution, we note that they differ in the correction term λ_i^2 in the inverse of $Z^T Z$. This reduces the bias in the solution caused by the noise.

The TLS solution can also be computed by using the singular value decomposition ([36], p. 503)

$$C^{(i)} = U\Sigma V^T,$$

where $U \in \mathbb{R}^{h \times h}$ and $V \in \mathbb{R}^{(n+2) \times (n+2)}$ are unitary matrices and Σ is a square diagonal matrix of dimension $k = \min(h, n+1)$, composed of the non-negative singular values of $C^{(i)}$ arranged in descending order along its main diagonal. The singular values are the positive square roots of the eigenvalues of $C^{(i)T} C^{(i)}$. Let $V = \begin{bmatrix} V_{*1} & \cdots & V_{*n} & V_{*(n+1)} & V_{*(n+2)} \end{bmatrix}$, where V_{*i} is the i-th column of V. Then, the solution is given by

$$\begin{pmatrix} \Theta_{*i}^{\text{TLS}} \\ -1 \end{pmatrix} = -\frac{V_{*(n+2)}}{V_{(n+2)(n+2)}}, \tag{7.28}$$

where $V_{(n+2)(n+2)}$ is the last element of $V_{*(n+2)}$. Numerically, this is a more robust method than computing the inverse of a matrix.

The improvement with respect to the standard LS is that the TLS approach allows us to consider uncertainty also on the regressors Z, not only on the dependent variables Y. Therefore, in the TLS the unknown parameters are optimised to minimise the deviation of the estimated model from both of these quantities.

7.5.2 Constrained total least squares

Unfortunately, the TLS solution is not optimal when the noise terms in Z and Y are correlated. Indeed, one of the main assumptions of the TLS is that the two noise terms are independent of each other. If there is some correlation between them, this knowledge can be used to improve the solution by employing the Constrained Total Least Squares (CTLS) technique, [35]. In the case of the problem in the form of Eq. (7.24), the two noise terms are obviously correlated because many elements of Z and Y_{*i} are coincident.

This prior information about the structure of $\Delta C^{(i)}$ can be explicitly taken into account in the optimisation. Let us first define a vector containing the minimal set of noise terms

$$\nu = \left(\nu_1(h) \;\cdots\; \nu_n(h) \;\cdots\; \nu_1(0) \;\cdots\; \nu_n(0)\right)^T \in \mathbb{R}^{n(h+1)}.$$

If ν is not white random noise, a whitening process using Cholesky factorisation is performed, [35]. Here, ν is assumed to be white noise and this whitening process is not necessary. The columns of $\Delta C^{(i)}$ can be written as

$$\Delta C^{(i)}_{\star j} = \left(\nu_j(h-1) \;\cdots\; \nu_j(0)\right)^T, \quad j = 1,\ldots,n,$$

$$\Delta C^{(i)}_{\star(n+1)} = 0_{h\times 1}, \quad \Delta C^{(i)}_{\star(n+2)} = \Delta Y_{\star j}. \tag{7.29}$$

It is possible to rewrite each column as $\Delta C^{(i)}_{\star j} = G^{(ij)} \nu$. To obtain an explicit form of the matrices $G^{(ij)}$, we first define the column vectors

$$e^{(j)} = (0 \cdots 0 \quad 1 \quad 0 \ldots 0)^T \in \mathbb{R}^n, \quad j = 1,\ldots,n,$$

containing all zero elements, except for the j-th element, which is equal to 1. We have

$$\Delta C^{(i)}_{\star j} = \left(\nu_j(h-1) \;\cdots\; \nu_j(0)\right)^T$$
$$= \left[\; 0_{h\times n} \quad (I_h \otimes e_j)^T \;\right] \nu$$

and hence

$$G^{(ij)} = \left[\; 0_{h\times n} \quad (I_h \otimes e_j)^T \;\right]$$

for $i = 1,\ldots,n$, where $\otimes$ denotes the Kronecker product. Also, from Eq. (7.29)

$$G^{(i(n+1))} = 0_{h\times n(h+1)},$$
$$G^{(i(n+2))} = \left[(I_h \otimes e_i)^T \quad 0_{h\times n}\right].$$

Since $\Delta C^{(i)}$ can be written as

$$\Delta C^{(i)} = \left(G^{(i1)}\nu \;\ldots\; G^{(i(n+2))}\nu\right),$$

then the TLS problem can be recast as

$$\min_{\nu,\Theta_{\star i}} \|\nu\|^2$$

$$\text{s.t.} \left[C^{(i)} + \left(G^{(i1)}\nu \;\ldots\; G^{(i(n+2))}\nu\right)\right] \begin{bmatrix} \Theta_{\star i} \\ -1 \end{bmatrix} = 0. \tag{7.30}$$

This is called the Constrained Total Least Squares (CTLS) problem. With the following definition:

$$H_\theta := \sum_{r=1}^{n} a_{ir}\, G_r + b_r\, G_{n+1} - G_{n+2} = \sum_{r=1}^{n+1} \Theta_{ri}\, G_r - G_{n+2}, \tag{7.31}$$

where Θ_{ri} for $r = 1, \ldots, n$ is the r-th element of the i-th row of A_d and $\Theta_{(n+1)i}$ is the i-th element of B_d, Eq. (7.30) can be written in the following form:

$$C^{(i)} \begin{bmatrix} \Theta_{*i} \\ -1 \end{bmatrix} + H_\theta \nu = 0.$$

Solving for ν, we get

$$\nu = -H_\theta^\dagger C^{(i)} \begin{bmatrix} \Theta_{*i} \\ -1 \end{bmatrix}, \tag{7.32}$$

where $H_\theta^\dagger$ is the pseudoinverse of H_θ. Hence, the original constrained minimisation problem, Eq. (7.30), is transformed into an unconstrained minimisation problem as follows:

$$\min_{\nu, \Theta_{*i}} \|\nu\|^2 = \min_{\Theta_{*i}} \begin{bmatrix} \Theta_{*i}^T & -1 \end{bmatrix} C^{(i)T} H_\theta^{\dagger T} H_\theta^\dagger C^{(i)} \begin{bmatrix} \Theta_{*i} \\ -1 \end{bmatrix}. \tag{7.33}$$

Now, we introduce two assumptions which make the formulation simpler.

1. The number of measurements is always strictly greater than the number of unknowns, i.e. we only consider the overdetermined case, explicitly $h + 1 > n + 2$, that is $h > n + 1$.

2. H_θ is full rank.

Then the pseudoinverse $H_\theta^\dagger$ is given by

$$H_\theta^\dagger = H_\theta^T \left(H_\theta H_\theta^T \right)^{-1}$$

and the unconstrained minimisation problem can be further simplified as follows:

$$\min_{\Theta_{*i}} \begin{bmatrix} \Theta_{*i}^T & -1 \end{bmatrix} C^{(i)T} \left(H_\theta H_\theta^T \right)^{-1} C^{(i)} \begin{bmatrix} \Theta_{*i} \\ -1 \end{bmatrix}. \tag{7.34}$$

The starting guess for Θ_{*i} used in the above optimisation problem is simply the value returned by the solution of the standard least squares problem.

The problem to be solved is to find the values of $n(n + 1)$ parameters of a linear model that yield the best fit to the observations in the least squares sense. Hence, as assumed above, if the number of observations is always strictly greater than the number of regression coefficients, that is $h > n + 1$, then the problem admits a unique globally optimal solution. In the other case, $h \leq n + 1$, the interpolation problem is under-determined, and thus there exist infinitely many values of the optimisation variables that equivalently fit the experimental measurements. In this case, as noted previously, expedients such as clustering or smoothing techniques and using a bottom-up approach can be adopted. In particular, the introduction of sign constraints on the optimisation variables, derived from qualitative prior knowledge of the network topology, will result in a significant reduction of the solution space.

7.6 Exploiting time-varying models

Linear time-invariant models will not always be able to effectively capture the dynamics of highly nonlinear biomolecular networks. For example, many biomolecular regulatory networks produce limit cycle dynamics, e.g. circadian rhythms, [37], cAMP oscillations in aggregating *Dictyostelium discoideum* cells, [38], or Ca^{2+} oscillations, [39]. Since such robust oscillatory dynamics cannot be produced by purely linear time-invariant systems, it is unlikely that the underlying network will be accurately identified using linear time-invariant models. As discussed previously, however, the use of nonlinear models in the network inference process almost always leads to ill-defined problem formulations which are not computationally tractable. A potential solution to this problem is to adopt linear *time-varying* systems as the model for inferring biomolecular networks, as proposed in [13]. Although linear time-varying systems are still in a linear form, they have a much richer range of dynamic responses than linear time-invariant ones. Hence, a wider range of time-series expression profiles, including oscillatory trajectories, can be approximated by such models.

Recall that the dynamics of most biomolecular regulatory networks arise from complex biochemical interactions which are nonlinear and can be written as

$$\frac{dx_i(t)}{dt} = f_i\left(x_1(t), \ldots, x_n(t)\right) \tag{7.35}$$

for $i = 1, \ldots, n$, where $f_i(\cdot)$ is a function that describes the dynamical interactions on $x_i(t)$ from $x_1(t), \ldots, x_n(t)$. If $f_i(\cdot)$ and $x_j(t)$ increase and decrease in a synchronous fashion, i.e., $f_i(\cdot)$ increases or decreases as $x_j(t)$ increases or decreases, it is said that $x_j(t)$ activates $x_i(t)$. On the other hand, if $f_i(\cdot)$ and $x_j(t)$ increase and decrease in an asynchronous fashion, i.e. $f_i(\cdot)$ increases or decreases as $x_j(t)$ decreases or increases, it is said that $x_j(t)$ inhibits $x_i(t)$. Consider the following p-number of experimental data points:

$$\tilde{x}_i(t_k) = x_i(t_k) + \nu_i(t_k) \tag{7.36}$$

for $i = 1, 2, \ldots, n-1, n$ and $k = 1, 2, \ldots, p-1, p$, where the measurement $\tilde{x}_i(t_k)$ is corrupted by some measurement noise $\nu_i(t_k)$ and t_k is the sampling time. Typically, in experiments the sampling interval $t_{k+1} - t_k$ is not necessarily the same for all k and the statistical properties of the noise are also generally unknown.

The estimation of $f_i(\cdot)$, which involves fitting the time profile of the states and finding the structure of the function, is an ill-posed problem. On the other hand, if the model is assumed to be linear time-invariant, taking the form

$$\frac{dx_i(t)}{dt} \approx \sum_{j=1}^{n} a_{ij}\, x_j(t) \tag{7.37}$$

for $i = 1, \ldots, n$, then the constant coefficients, a_{ij}, may be estimated and the problem is well posed. In this case, however, the linear model may not be a good fit for the experimental data, which has been generated from nonlinear network interactions. A typical nonlinear phenomenon that cannot be approximated by a linear time-invariant model is a limit cycle. To render the estimation problem well posed while preserving the ability of the candidate model to closely fit the data, one can use the linear time-varying model

$$\frac{dx_i(t)}{dt} \approx \sum_{j=1}^{n} a_{ij}(t) x_j(t) \tag{7.38}$$

for $i = 1, \ldots, n$, where $a_{ij}(t)$ is a time-varying function. The estimation problem can be further simplified by limiting the rate of change of $a_{ij}(t)$ with time. This is reasonable, since the measurement frequency of any biological experiment is limited and therefore only information up to a certain frequency in the data can be correctly uncovered from the measurements. In this case, $a_{ij}(t)$ can be written as a finite sum of Fourier series, [40]

$$a_{ij}(t) = \alpha_{ij} \sin (\omega t + \phi_{ij}) + \beta_{ij} \tag{7.39}$$

where α_{ij}, ω, ϕ_{ij} and β_{ij} are the constants to be determined. β_{ij} represents the linear part of the interactions and the sinusoidal term approximates any nonlinear terms in the interactions. If needed, more sinusoidal terms can easily be included to more closely approximate the nonlinearities, at the cost of increasing the computational burden for the optimisation algorithm. By using the linear time-varying model, the following optimisation problem can be formulated:

$$\min_{\alpha_{ij}, \beta_{ij}, \phi_{ij}, \omega, x_i(t_1)} J_i = \frac{1}{\max_{t_k} |\hat{x}_i(t_k)|} \sum_{k=1}^{p} [x_i(t_k) - \hat{x}_i(t_k)]^2 \tag{7.40}$$

for $i = 1, 2, \ldots, n-1, n$, subject to Eq. (7.38), where $\hat{x}_i(t_k)$ is a numerically perturbed measurement, and $x_i(t)$ is the solution of Eq. (7.38). For a fixed i, the number of parameters to be estimated is $3n + 2$, including the initial condition of Eq. (7.38), $x_i(t_1)$. The appropriate choice of optimisation algorithm to solve the above problem depends on the number of parameters to be estimated — for small scale problems, the simplex search method implemented in MATLAB®, [41], may be used as in [13], while for larger scale problems randomisation based optimisation algorithms would be more appropriate, e.g., the simultaneous perturbation method in [42].

Note that the cost function is normalised by the maximum value of the measurements of each state. The optimisation problem is formulated separately for each $dx_i(t)/dt$ in order to reduce the rate of increase in the number of parameters to be estimated as the dimension of $x_i(t)$ increases. If the problem is formulated for all $x_i(t)$, the number of parameters increases according to $3n^2 + n + 1$. The price to be paid for reducing the number of parameters to be computed in this way is the increased effect of noise, since in order to solve Eq. (7.38) all the states except $x_i(t)$ have to be interpolated from the measurements and this will necessarily introduce the direct effect of noise on the estimate. To reduce this effect some elements of the noise could be filtered out before the data are used in the interpolation, e.g. by using Principal Component Analysis, [43], or the noisy measurements could be replaced by the solution of the differential equation (7.38). After the optimal solution is obtained for Eq. (7.40), if the optimal cost is smaller than a certain bound, for example 10% of the maximum of the measurements, the measurements are replaced by the solution of the differential equations, under the assumption that the model gives less noisy data without deteriorating the original measurements significantly. As the optimisation problem is solved from x_1 to x_n, more measurements may be replaced. At the final stage, all measurements except the measurements for x_n could be replaced by the filtered states. To remove the unbalanced noise effect, the same procedure is repeated in the opposite direction, i.e., starting from x_{n-1} to x_1 since the earlier states may be affected more by the noise.

The problem formulation of Eq. (7.40) is a very flexible one, since it can cope with cases where the sampling time is not evenly distributed, and weighting can also be used in the cost when the error bar at each sampling time is different. The optimisation problem is, however, nonlinear, and hence it may have many local solutions. If biologically plausible ranges for the parameters are known, these can be used in choosing a better initial guess for the optimisation, in order to improve the chances of finding the globally optimal solution. Initial values for the parameters may also be chosen by inspecting the finite difference magnitude of $\tilde{x}_i(t_k)$, since, although the data are corrupted by noise, the rate of change of $x_i(t)$ should still not be very different from the magnitude of the finite difference. The initial guess for ω can be obtained by calculating the dominant frequency of the measurement data using Fourier transforms, [40]. Results of the application of an inference algorithm based on the aboveapproach to a number of different biological examples are described in [13].

7.7 Case Study XI: Inferring regulatory interactions in the innate immune system from noisy measurements

Biology background: All organisms are constantly being exposed to infectious agents and yet, in most cases, they are able to resist these infections due to the action of their immune systems. The immune system is composed of two major subdivisions, the innate or non-specific immune system and the adaptive or specific immune system. The innate immune system is made up of the cells and mechanisms that mediate a non-specific defense of the host from infection by other organisms. In contrast to the adaptive immune system, the innate immune system recognises and responds to pathogens in a generic way and does not confer long-lasting or protective immunity on the host. The innate immune system thus provides the first line of defense against infection and is found in all classes of plant and animal life. From an evolutionary perspective, it is believed to be an early form of defense strategy, and indeed it is the dominant immune system found in plants, fungi, insects and in primitive multicellular organisms.

The main function of the immune system is to distinguish between self and non-self, in order to protect the organism from invading pathogens and to eliminate modified or altered cells (e.g. malignant cells). Since pathogens may replicate intracellularly (viruses and some bacteria and parasites) or extracellularly (most bacteria, fungi and parasites), different components of the immune system have evolved to protect against these different types of pathogens. Infection with an organism does not necessarily lead to diseases, since the immune system in most cases will be able to eliminate the infection before disease occurs. Disease occurs only when the bolus of infection is high, when the virulence of the invading organism is great or when immunity is compromised. Although the immune system, for the most part, has beneficial effects, there can be detrimental effects as well. During inflammation, which is the response to an invading organism, there may be local discomfort and collateral damage to healthy tissue as a result of the toxic products produced by the immune response. In addition, in some cases the immune response can be directed toward self tissues resulting in autoimmune disease.

The innate immune system functions by recruiting immune cells to sites of infection, through the production of chemical factors called cytokines that are secreted by macrophages. Cytokines are specialised regulatory proteins, such as the interleukins and lymphokines, that are released by cells of the immune system and act as intercellular mediators in the generation of an immune response.

Other functions of the innate immune response include the clearance of dead cells or antibody complexes, the identification and removal of foreign substances present in organs, tissues, the blood and lymph nodes, by specialised white blood cells, and the activation of the adaptive immune system through a process known as antigen presentation.

Proper regulation of the innate immune system is crucial for host survival, and breakdown of the immune system regulatory mechanisms can lead to inflammatory disease. It therefore comes as no surprise that the control mechanisms employed in nature to regulate the immune system response are extraordinarily complex, making them a prime candidate for investigation using systems biology approaches.

In [44], the authors used cluster analysis of a comprehensive set of transcriptomic data derived from Toll-like receptor (TLR)-activated macrophages to identify a prominent group of genes that appear to be regulated by activating transcription factor 3 (ATF3), a member of the CREB/ATF family of transcription factors. Network analysis predicted that ATF3 is part of a transcriptional complex that also contains members of the nuclear factor (NF)-κB family of transcription factors. Promoter analysis of the putative ATF3-regulated gene cluster demonstrated an over-representation of closely apposed ATF3 and NF-κB binding sites, which was verified by chromatin immunoprecipitation and hybridisation to a DNA microarray. This cluster included important cytokines such as interleukin (IL)-6 and IL-12b. ATF3 and Rel (a component of NF-κB) were shown to bind to the regulatory regions of these genes upon macrophage activation. Thus, the biochemical network through which interleukin (IL)-6 and IL-12b interact with activating transcription factor 3 (ATF3) and Rel (a component of NF-κB) appears to form an important part of the innate immune system response. In [44], a kinetic model for the expression of *IL6* mRNA by *ATF3* and *Rel* was proposed as follows:

$$\frac{d(\mathit{Il6})}{dt} = -\frac{1}{\tau}\mathit{Il6} + \frac{1}{\tau\left(1 + e^{-\beta_{\mathrm{Rel}}\mathit{Rel} - \beta_{\mathrm{ATF3}}\mathit{ATF3}}\right)} \tag{7.41}$$

where $\tau = 600/\ln(2)$, $\beta_{\mathrm{Rel}} = 7.8$ and $\beta_{\mathrm{ATF3}} = -4.9$. This kinetic model was developed to match the experimental data shown in Fig. 7.8. Similarly, a kinetic model for *IL12* is given by

$$\frac{d(\mathit{Il12})}{dt} = -\frac{1}{\tau}\mathit{Il12} + \frac{1}{\tau\left(1 + e^{-\beta_{\mathrm{Rel}}\mathit{Rel} - \beta_{\mathrm{ATF3}}\mathit{ATF3}}\right)} \tag{7.42}$$

where $\tau = 600/\ln(2)$, $\beta_{\mathrm{Rel}} = 18.5$, and $\beta_{\mathrm{ATF3}} = -9.6$. We now consider the problem of estimating A, the Jacobian matrix of $f(x)$ for this system, from the noisy experimental data given in Fig. 7.8, [45]. Using the proposed kinetic models, an analytical expression for one row of A can be obtained for *Il6* as

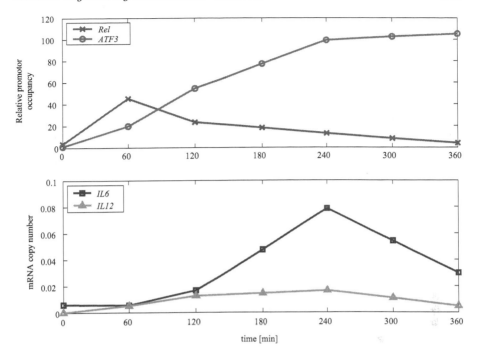

FIGURE 7.8: The measurements of Rel, ATF3, Il6 and Il12 are taken from [44]. The actual data in [44] are measured at 0, 60, 120, 240 and 360 minutes. To make the measurements equally spaced in time, the data shown at 180 and 300 minutes are interpolated.

follows:

$$\frac{\partial(dIl6/dt)}{\partial Il6} = -\frac{1}{\tau} = -\frac{\ln(2)}{600} \approx -0.00116 \tag{7.43a}$$

$$\frac{\partial(dIl6/dt)}{\partial Rel} = \frac{-\beta_{\text{Rel}}e^{-\beta_{\text{Rel}}Rel-\beta_{\text{ATF3}}ATF3}}{\tau\left(1 + e^{-\beta_{\text{Rel}}Rel-\beta_{\text{ATF3}}ATF3}\right)^2} \tag{7.43b}$$

$$\frac{\partial(dIl6/dt)}{\partial ATF3} = \frac{-\beta_{\text{ATF3}}e^{-\beta_{\text{Rel}}Rel-\beta_{\text{ATF3}}ATF3}}{\tau\left(1 + e^{-\beta_{\text{Rel}}Rel-\beta_{\text{ATF3}}ATF3}\right)^2} \tag{7.43c}$$

and a similar result can be obtained for *Il12*. Unfortunately, the second and the third partial derivatives above cannot be calculated unless the equilibrium condition values for *Rel* and *ATF3* are known. However, we can obtain the following ratio:

$$\frac{\partial(dIl6/dt)}{\partial Rel}\left[\frac{\partial(dIl6/dt)}{\partial ATF3}\right]^{-1} = \frac{\partial ATF3}{\partial Rel} = \frac{\beta_{\text{Rel}}}{\beta_{\text{ATF3}}} = \frac{7.8}{-4.9} = -1.59 \tag{7.44}$$

and therefore we can partially validate the Jacobian estimation from the data against the proposed model by checking the value of this ratio. The equivalent ratio for the case of *Il12* is -1.93. Note that the negative sign of this value is crucial, since it predicts that ATF3 is a negative regulator of Il6 and Il12b transcription, a hypothesis which was subsequently validated using Atf3-null mice in [44]. ATF3 seems to inhibit Il6 and Il12b transcription by altering the chromatin structure, thereby restricting access to transcription factors. Because ATF3 is itself induced by lipopolysaccharide, it seems to regulate TLR-stimulated inflammatory responses as part of a negative feedback loop.

To obtain the data shown in Fig. 7.8, wild type mice were stimulated (or perturbed) by 10 ng ml^{-1} lipopolysaccharide (LPS). The data were sampled at intervals of 10 minutes but the original data at 180 and 300 minutes were not given; hence, they are interpolated for our study to make all data equally spaced in time. Naturally, the measurement data will include significant amounts of noise, and thus we expect that the direct calculation of the Jacobian using the conventional least squares algorithm may produce biased or inaccurate results. Note that since the number of states is 3, the number of perturbations is 1 and the number of data points for each state is 7, there is relatively little data with which to accurately estimate the Jacobian for this particular example. In addition, since the equilibrium point is not given, the measurements we have are not relative measurements $\Delta \tilde{x}_k$ but absolute measurements $\tilde{x}_k$. This presents no difficulty, however, since the problem formulation to estimate the Jacobian using $\tilde{x}_k$ is exactly the same as the one for $\Delta \tilde{x}_k$ — see [46] for more details.

To investigate the effect of measurement noise on the quality of the inference results, each of the three different least squares algorithms described above (LS, TLS and CTLS) were applied to this problem. For *Il6* the key result obtained is that the standard least squares algorithm gives the wrong (positive) sign for the ratio defined above, whereas the more advanced algorithms give the correct sign. The correct ratio of *Rel* and *ATF3* to *Il6* is -1.59 and the estimated values computed with the LS, TLS and CTLS algorithms are 1.43, -3.73, and -6.35, respectively. Thus, only by using the TLS or CTLS algorithms can the negative regulation effect of ATF3 be confirmed from the noisy data shown in Fig. 7.8. For *Il12*, the ratio calculated from each method, i.e., LS, TLS and CTLS, is -4.53, -2.46, and -1.98, respectively. Therefore, in this case all three algorithms predict the negative regulation role of ATF3 correctly. However, the ratio computed from the CTLS, -1.98, is by far the closest to the true value (-1.93) predicted by the model.

7.8 Case Study XII: Reverse engineering a cell cycle regulatory subnetwork of *Saccharomyces cerevisiae* from experimental microarray data

Biology background: The cell cycle is a series of events that takes place inside a cell leading to its division and replication. In prokaryotic cells, the cell cycle occurs via a process called binary fission. In eukaryotic cells, the cell cycle consists of four distinct phases: the first three, G1 phase, S phase (synthesis) and G2 phase, are collectively known as the interphase, during which the cell grows and accumulates the nutrients needed for mitosis and duplication of its DNA. The fourth phase is termed the M phase (mitosis), which is itself made up of two tightly coupled processes: mitosis, in which the cell's chromosomes are divided between the two daughter cells and cytokinesis, in which the cell's cytoplasm divides in half, forming distinct cells. Activation of each phase is dependent on the proper progression and completion of the previous one. Cells that have temporarily or reversibly stopped dividing are said to have entered a state of quiescence called G0 phase. The cell cycle is a fundamental developmental process in biology, in which a single-celled fertilised egg grows into a mature organism. It is also the process by which hair, skin, blood cells and many internal organs are renewed.

Correct regulation and control of the cell cycle is crucial to the survival of a cell and requires the detection and repair of genetic damage as well as the prevention of uncontrolled cell division. Progress through the cell cycle is controlled by two key classes of regulatory molecules, cyclins and cyclin-dependent kinases (CDKs), and takes place in an sequential and directional manner which cannot be reversed. Many of the genes encoding cyclins and CDKs are conserved among all eukaryotes, but in general more complex organisms have more elaborate cell cycle control systems that incorporate more individual components. Many of the key regulatory genes were first identified by studies of the yeast *Saccharomyces cerevisiae*, where it appears that a semi-autonomous transcriptional network acts along with the CDK-cyclin machinery to regulate the cell cycle. Several gene expression studies have identified approximately 800 to 1200 genes that change expression over the course of the cell cycle — they are transcribed at high levels at specific points in the cycle and remain at lower levels throughout the rest of it. While the set of identified genes differs between studies due to the computational methods and criterion used to identify them, each study indicates that a large portion of yeast genes are temporally regulated.

Disruption of the cell cycle can lead to cancer. Mutations in cell cycle inhibitor genes such as p53 can cause the cell to multiply uncontrollably, resulting in the formation of a tumour. Although the duration of the cell cycle in tumour cells is approximately the same as that of normal cells, the proportion of cells that are actively dividing (versus the number in the quiescent G0 phase) is much higher. This results in a significant increase in cell numbers as the number of cells that die by apoptosis or senescence remains constant. Many cancer therapies specifically target cells which are actively undergoing a cell cycle, since in these cells the DNA is relatively exposed and hence susceptible to the action of drugs or radiation. One process, known as debulking, removes a significant mass of the tumour, which leads a large number of the remaining tumour cells to change from the G0 to G1 phase due to increased availability of nutrients, oxygen, growth factors, etc. These cells which have just entered the cell cycle are then targeted for destruction using radiation or chemotherapy.

In this Case Study, we illustrate the application of the PACTLS algorithm to the problem of reverse engineering a regulatory subnetwork of the cell cycle in *Saccharomyces cerevisiae* from experimental microarray data. The network is based on the model proposed by [47] for transcriptional regulation of cyclin and cyclin/CDK regulators and the model proposed by [48], where the main regulatory circuits that drive the gene expression program during the budding yeast cell cycle are considered. The network is composed of 27 genes: 10 genes that encode for transcription factor proteins (ace2, fkh1, swi4, swi5, mbp1, swi6, mcm1, fkh2, ndd1, yox1) and 17 genes that encode for cyclin and cyclin/CDK regulatory proteins (cln1, cln2, cln3, cdc20, clb1, clb2, clb4, clb5, clb6, sic1, far1, spo12, apc1, tem1, gin4, swe1 and whi5). The microarray data have been taken from [49], selecting the data set produced by the alfa factor arrest method. Thus, the raw data set consists of $n = 27$ genes and 18 data points. A smoothing algorithm has been applied in order to filter the measurement noise and to increase by interpolation the number of observations. The gold standard regulatory network comprising the chosen 27 genes has been drawn from the BioGRID database, [50], taking into account the information of [47] and [48]: the network consists of 119 interactions, not including the self-loops, yielding a value of the sparsity coefficient, defined by $\eta = 1 - \#\text{edges}/(n^2 - n)$, equal to 0.87.

7.8.1 PACTLS: An algorithm for reverse engineering partially known networks from noisy data

In this subsection, we describe the PACTLS algorithm, [51], a method devised for the reverse engineering of partially known networks from noisy data. PACTLS uses the CTLS technique to optimally reduce the effects of measurement noise in the data on the reliability of the inference results, while exploiting qualitative prior knowledge about the network interactions with an edge selection heuristic based on mechanisms underpinning scale-free network generation, i.e. network growth and preferential attachment (PA).

The algorithm allows prior knowledge about the network topology to be taken into account within the CTLS optimisation procedure. Since each element of A can be interpreted as the weight of the edge between two nodes of the network, this goal can be achieved by constraining some of the optimisation variables to be zero and others to be strictly positive (or negative), and using a constrained optimisation problem solver, e.g. the nonlinear optimisation function *fmincon* from the MATLAB® Optimisation toolbox, to solve Eq. (7.34). Similarly, we can impose a sign constraint on the i-th element of the input vector, b_i, if we *a priori* know the qualitative (i.e. promoting or repressing) effect of the perturbation on the i-th node. Alternatively, an edge can be easily pruned from the network by setting to zero the corresponding entry in the minimisation problem.

So far we have described a method to add/remove edges and to introduce constraints on the sign of the associated weights in the optimisation problem. The problem remains of how to devise an effective strategy to select the nonzero entries of the connectivity matrix.

The initialisation network for the devised algorithm has only self-loops on every node, which means that the evolution of the i-th state variable is always influenced by its current value. This yields a diagonal initialisation matrix, $\hat{A}^{(0)}$. Subsequently, new edges are added step by step to the network according to the following iterative procedure:

P1) A first matrix, $\bar{A}$, is computed by solving the optimisation problem (7.34) for each row, without setting any optimisation variable to zero. The available prior information is taken into account at this point by adding the proper sign constraints on the corresponding entries of A before solving the optimisation problem. Since it typically exhibits all nonzero entries, matrix $\bar{A}$ is not representative of the network topology, but is rather used to weight the relative influence of each entry on the system's dynamics. This information will be used to select the edges to be added to the network at each step. Each element of $\bar{A}$ is normalised with respect to the values of the other elements in the same row and column, which yields the matrix $\tilde{A}$, whose elements are defined as

$$\tilde{A}_{ij} = \frac{\bar{A}_{ij}}{\left(\|\bar{A}_{\star,j}\| \cdot \|\bar{A}_{i,\star}\|\right)^{1/2}}.$$

P2) At the k-th iteration, the edges ranking matrix $\tilde{G}^{(k)}$ is computed:

$$\tilde{G}_{ij}^{(k)} = \frac{|\tilde{A}_{ij}|p_j^{(k)}}{\sum_{l=1}^{n} p_l^{(k)} |\tilde{A}_{il}|}, \qquad (7.45)$$

where

$$p_j^{(k)} = \frac{K_j^{(k)}}{\displaystyle\sum_{l=1}^{n} K_l^{(k)}} \tag{7.46}$$

is the probability of inserting a new edge starting from node j and $K_l^{(k)}$ is the number of outgoing connections from the l-th node at the k-th iteration. The $\mu(k)$ edges with the largest scores in $\tilde{G}^{(k)}$ are selected and added to the network; $\mu(\cdot)$ is chosen as a decreasing function of k, that is $\mu(k) = \lceil n/k \rceil$. Thus, the network grows rapidly at the beginning and is subsequently refined by adding smaller numbers of nodes at each iteration. The form of the function $p(\cdot)$ stems from the so-called *preferential attachment* (PA) mechanism, which states that in a growing network new edges preferentially start from *popular* nodes (those with the highest connectivity degree, i.e. the hubs). By exploiting the mechanisms of network growth and PA, we are able to guide the network reconstruction algorithm to increase the probability of producing a network with a small number of hubs and many poorly connected nodes. Note also that, for each edge, the probability of incidence is blended with the edge's weight estimated at point P1; therefore, the edges with larger estimated weights have a higher chance to be selected. This ensures that the interactions exerting greater influence on the network dynamics have a higher probability of being selected.

P3) The structure of nonzero elements of $\hat{A}^{(k)}$ is defined by adding the entries selected at point P2 to those selected up to iteration $k-1$ (including those derived by *a priori* information), and the set of inequality constraints is updated accordingly; then Problem 7.34 for each row, with the additional constraints, is solved to compute $\hat{A}^{(k)}$.

P4) The residuals generated by the identified model are compared with the values obtained at the previous iterations; if the norm of the vector of residuals has decreased, in the last two iterations, at least by a factor ϵ_r with respect to the value at the first iteration, then the procedure iterates from point P2; otherwise it stops and returns the topology described by the sparsity pattern of $\hat{A}^{(k-2)}$. The factor ϵ_r is inversely correlated with the number of edges inferred by the algorithm; on the other hand, using a smaller value of ϵ_r raises the probability of obtaining false positives. By conducting numerical tests for different values of ϵ_r, we have found that setting $\epsilon_r = 0.1$ yields a good balance between the various performance indices.

Concerning the input vector, we assume that the perturbation targets and the qualitative effects of the perturbation are known; thus, the pattern (but not the values of the nonzero elements) of $\hat{B}$ is preassigned at the initial step and the corresponding constraints are imposed in all the subsequent iterations.

7.8.2 Results

Fig. 7.9 shows the results obtained by PACTLS assuming four different levels of prior knowledge (PK) from 10% to 40% of the network. The performance is evaluated by using two common statistical indices (see [52], p. 138):

- *Sensitivity* (Sn), defined as

$$Sn = \frac{TP}{TP + FN},$$

  which is the fraction of actually existing interactions (TP:=true positives, FN:=false negatives) that the algorithm infers, also termed *Recall*, and

- *Positive Predictive Value* (PPV),

$$PPV = \frac{TP}{TP + FP},$$

  which measures the reliability of the interactions (FP:=false positives) inferred by the algorithm, also named *Precision*.

To compute these performance indexes, the weight of an edge is not considered, but only its existence, so the network is considered as a directed graph. The performance of PACTLS is compared with one of the most popular statistical methods for network inference, dynamic Bayesian networks. For these purposes we used the software BANJO (BAyesian Network inference with Java Objects), [53], which performs network structure inference for static and dynamic Bayesian networks (DBNs).

The performance of both approaches is compared in Fig. 7.9. In order to further validate the inference capability of the algorithms, the figure shows also the results obtained by a random selection of the edges, based on a binomial distribution: given any ordered pair of nodes, the existence of a directed edge between them is assumed true with probability p_r and false with probability $1 - p_r$. By varying the parameter p_r in $[0, 1]$, the random inference algorithm produces results shown as the solid curves on the (PPV, Sn) plot in Fig. 7.9.

The performance of PACTLS is consistently significantly better than the method based on DBNs: the distance of the PACTLS results from the random curve is almost always larger than those obtained with the BANJO software, which is not able to achieve significant Sn levels, probably due to the low number of time points available. Moreover, the results show that the performance of PACTLS improves progressively when the level of prior knowledge increases. Fig. 7.10 shows the regulatory subnetwork inferred by CORE–Net, assuming 50% of the edges are *a priori* known. Seven functional interactions, which are present in the gold standard network, have been correctly inferred. Moreover, seven other functional interactions have been returned which are not present in the gold standard network. To understand if the latter should

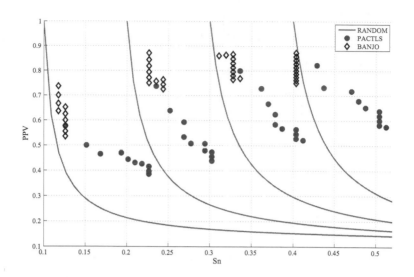

FIGURE 7.9: Results for the cell cycle regulatory subnetwork of *Saccharomyces cerevisiae* assuming different levels of prior knowledge (PK=10, 20, 30, 40%).

be classified as TP or FP, we manually mined the literature and the biological databases and uncovered the following results:

- The interaction between *mbp1* and *gin4* is reported by the YEAS-TRACT database [54]: *mbp1* is reported to be a transcription factor for *gin4*;

- A possible interaction between *fkh2* and *swi6* is also reported by the YEASTRACT database: *fkh2* is reported to be a potential transcription factor for *swi6*);

- The interaction between *clb1* and *swi5* appears in Fig. 1 in [48], where the scheme of the main regulatory circuits of the budding yeast cell cycle is described.

Thus, these three interactions can be classified as TP as well and are reported as light-grey dashed edges in Fig. 7.10.

Concerning the other inferred interactions, two of them can be explained by the indirect influence of *swi6* on *fkh1* and *fkh2*, which is mediated by *ndd1*: in fact, the complexes SBF (Swi4p/Swi6p) and MBF (Mbp1p/Swi6p) both regulate *ndd1*, [47], which can have a physical and genetic interaction with *fkh2*. Moreover, *fkh1* and *fkh2* are forkhead family transcription factors

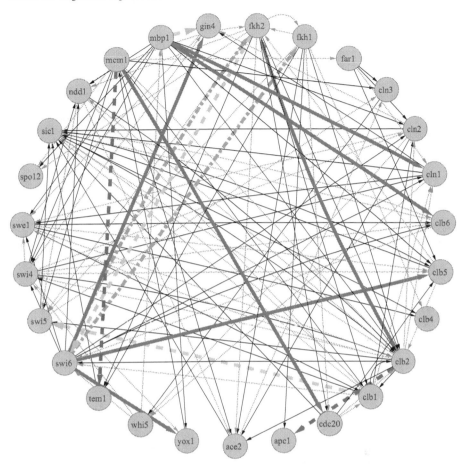

FIGURE 7.10: Gene regulatory subnetwork of *S. cerevisiae* inferred by CORE–Net with 50% of the edges *a priori* known (thin solid edges). Results according to the gold standard network drawn from the BioGRID database: TP=thick solid edge, FN=dotted edge, FP=thick dashed and dashed-dotted edge. The thick light-grey dashed edges are not present in the BioGRID database; however, they can be classified as TP according to other sources. The FP thick dashed-dotted edges are indirect interactions mediated by *ndd1*. No information has been found regarding the interactions denoted by the thick dark-grey dashed edges.

which positively influence the expression of each other. Thus, the inferred interactions are not actually between adjacent nodes of the networks and have to be formally classified as FP (these are reported as dashed-dotted edges in Fig. 7.10).

Concerning the last two interactions, that is $clb2 \rightarrow apc1$ and $mcm1 \rightarrow tem1$, since we have not found any information on them in the literature, in the absence of further experimental evidence they have to be classified as FP (reported as dark-grey dashed edges in Fig. 7.10).

The results obtained in this Case Study highlight the potential of the approaches described in this chapter for reverse engineering biomolecular networks, and in particular confirm the importance of dealing with measurement noise and exploiting prior knowledge to improve the reliability of the network inference.

References

[1] Brown PA and Botstein D. Exploring the new world of the genome with DNA microarrays. *Nature Genetics*, 21(1):33–37, 1999.

[2] Lipschutz RJ, Fodor SPA, Gingeras TR, and Lockhart DJ. High density synthetic oligonucleotide arrays. *Nature Genetics*, 21(1):20–24, 1999.

[3] Balding DJ, Stumpf M, and Girolami M. (Editors). *Handbook of Statistical Systems Biology*. Chichester: Wiley, 2011.

[4] d'Alche-Buc F and Schachter V. Modeling and simulation of biological networks. In *Proceedings of the International Symposium on Applied Stochastic Models and Data Analysis (ASMDA'05)*, Brest, France, May 2005.

[5] Hecker M, Lambeck S, Toepfer S, van Someren E, and Guthke R. Gene regulatory network inference: data integration in dynamic models — a review. *BioSystems*, 96:86–103, 2009.

[6] Gardner TS, di Bernardo D, Lorenz D, and Collins JJ. Inferring genetic networks and identifying compound mode of action via expression profiling. *Science*, 301:102–105, 2003.

[7] di Bernardo D, Gardner TS, and Collins JJ. Robust identification of large genetic networks. In *Proceedings of the Pacific Symposium on Biocomputing (PSB'04)*, Waimea, USA, January 2004.

[8] Cho K-H, Kim JR, Baek S, Choi H-S, and Choo S-M. Inferring biomolecular regulatory networks from phase portraits of time-series expression profiles. *FEBS Letters*, 580(14):3511–3518, 2006.

[9] Kim S, Kim J, and Cho K-H. Inferring gene regulatory networks from temporal expression profiles under time-delay and noise. *Computational Biology and Chemistry*, 31(4):239–245, 2007.

[10] Han S, Yoon Y, and Cho K-H. Inferring biomolecular interaction networks based on convex optimization. *Computational Biology and Chemistry*, 31(5-6):347–354, 2007.

[11] Montefusco F, Cosentino C, and Amato F. CORE–Net: Exploiting prior knowledge and preferential attachment to infer biological interaction networks. *IET Systems Biology*, 4(5):296–310, 2010.

[12] Guthke R, Moller U, Hoffman M, Thies F, and Topfer S. Dynamic network reconstruction from gene expression data applied to immune response during bacterial infection. *Bioinformatics*, 21:1626–1634, 2005.

[13] Kim J, Bates DG, Postlethwaite I, Heslop-Harrison P, and Cho K-H. Linear time-varying models can reveal non-linear interactions of biomolecular regulatory networks using multiple time-series data. *Bioinformatics*, 24:1286–1292, 2008.

[14] Franklin GF, Powell JD, and Emani-Naeini A. *Feedback Control of Dynamic Systems*. Boston: Addison-Wesley Publishing Company Inc., 3rd edition, 1994.

[15] Meyer CD. *Matrix Analysis and Applied Linear Algebra*. Philadelphia, PA: SIAM Press, 2000.

[16] di Bernardo D, et al. Chemogenomic profiling on a genomewide scale using reverse-engineered gene networks. *Nature Biotechnology*, 23:377–383, 2005.

[17] Bansal M, Della Gatta G, and di Bernardo D. Inference of gene regulatory networks and compound mode of action from time course gene expression profiles. *Bioinformatics*, 22(7):815–822, 2006.

[18] Bansal M and di Bernardo D. Inference of gene networks from temporal gene expression profiles. *IET Systems Biology*, 1(5):306–312, 2007.

[19] Bansal M, Belcastro V, Ambesi-Impiombato A, and di Bernardo D. How to infer gene regulatory networks from expression profiles. *Molecular Systems Biology*, 3:78, 2007.

[20] Ljung L. *System Identification: Theory for the User*. Upper Saddle River: Prentice Hall, 1999.

[21] Bonneau R, Reiss DJ, Shannon P, Facciotti M, Hood L, Baliga NS, and Thorsson V. The Inferelator: an algorithm for learning parsimonious regulatory networks from systems-biology data sets de novo. *Genome Biology*, 7:R36, 2006.

[22] Cosentino C, Curatola W, Montefusco F, Bansal M, di Bernardo D, and Amato F. Linear matrix inequalities approach to reconstruction of biological networks. *IET Systems Biology*, 1(3):164–173, 2007.

[23] Julius A, Zavlanos M, Boyd S, and Pappas GJ. Genetic network identification using convex programming. *IET Systems Biology*, 3(3):155–166, 2009.

[24] Zhou K. *Essentials of Robust Control*. Upper Saddle River: Prentice Hall, 1998.

[25] Boyd S, El Ghaoui L, Feron E, and Balakrishnan V. *Linear Matrix Inequalities in System and Control Theory*. Philadelphia, PA: SIAM Press, 1994.

[26] Gahinet P, Nemirovski A, Laub AJ, and Chilali M. *LMI Control Toolbox*, Natick: The Mathworks, 1995.

[27] Albert R and Barabasi A-L. Topology of evolving networks: local events and universality. *Physical Review Letters*, 85(24):5234–5237, 2000.

[28] Jeong H, Tombor B, Albert R, Oltvai ZN, and Barabasi A-L. The large-scale organization of metabolic networks. *Nature*, 407:651–654, 2000.

[29] Wagner A. The yeast protein interaction network evolves rapidly and contains few redundant duplicate genes. *Molecular Biology and Evolution*, 18:1283–1292, 2001.

[30] Featherstone DE and Broadie K. Wrestling with pleiotropy: genomic and topological analysis of the yeast gene expression network. *Bioessays*, 24:267–274, 2002.

[31] Luscombe NM, Qian J, Zhang Z, Johnson T, and Gerstein M. The dominance of the population by a selected few: power-law behaviour applies to a wide variety of genomic properties. *Genome Biology*, 3(8):research0040.10040.7, 2002.

[32] Golub GH and Van Loan CF. An analysis of the total least squares problem. *SIAM Journal on Numerical Analysis*, 17(6):883–893, 1980.

[33] Van Huffel S and Vandewalle J. The total least squares problem: computational aspects and analysis. *Frontiers in Applied Mathematics Series* 9, Philadelphia, PA: SIAM, 1991.

[34] Abatzoglou T and Mendel J. Constrained total least squares. In *Proceedings of the IEEE International Conference on Acoustics, Speech and Signal Processing*, 12:1485–1488, 1987.

[35] Abatzoglou TJ, Mendel JM, and Harada GA. The constrained total least squares technique and its application to harmonic superresolution. *IEEE Transactions on Signal Processing*, 39(5):1070–1087, 1991.

[36] Skogestad S and Postlethwaite I. *Multivariable Feedback Control: Analysis and Design*. Chichester: Wiley, 1996.

[37] Goldbeter, A. A model for circadian oscillations in the *Drosophila* period protein (PER). *Proceedings of the Royal Society B: Biological Sciences*, 261:319–324, 1995.

[38] Laub MT and Loomis WF. A molecular network that produces spontaneous oscillations in excitable cells of *Dictyostelium*. *Molecular Biology of the Cell*, 9:3521–3532, 1998.

[39] Dellen BK, Barber MJ, Ristig ML, Hescheler J, Sauer H, and Wartenberg M. Ca^{2+} oscillations in a model of energy-dependent Ca^{2+} uptake by the endoplasmic reticulum. *Journal of Theoretical Biology*, 237:279–290, 2005.

[40] Proakis JG and Manolakis DK. *Digital Signal Processing: Principles, Algorithms, and Applications*. 4th Edition, Upper Saddle River: Prentice Hall, 2006.

[41] *MATLAB 7 Function Reference: Volume 2 (F-O)*. Natick: Mathworks, Inc. 2007.

[42] Spall, JC. *Introduction to Stochastic Search and Optimization: Estimation, Simulation, and Control*. Chichester: Wiley, 2003.

[43] Sanguinetti G, Milo M, Rattray M, and Lawrence ND. Accounting for probe-level noise in principal component analysis of microarray data. *Bioinformatics*, 21:3748–3754, 2005.

[44] Gilchrist M, Thorsson V, Li B, Rust AG, Korb M, Kennedy K, Hai T, Bolouri H, and Aderem A. Systems biology approaches identify ATF3 as a negative regulator of Toll-like recepter 4. *Nature*, 441(11):173–178, 2006.

[45] Kim J, Bates DG, Postlethwaite I, Heslop-Harrison P, and Cho K-H. Least-squares methods for identifying biochemical regulatory networks from noisy measurements. *BMC Bioinformatics* 8:8, 2007.

[46] Schmidt H, Cho K-H, and Jacobsen EW. Identification of small scale biochemical networks based on general type system perturbations. *FEBS Journal*, 272:2141–2151, 2005.

[47] Simon I, Barnett J, Hannett N, Harbison CT, Rinaldi NJ, Volkert TL, Wyrick JJ, Zeitlinger J, Gifford DK, Jaakkola TS, and Young RA. Serial regulation of transcriptional regulators in the yeast cell cycle. *Cell*, 106(1):697–708, 2001.

[48] Bahler J. Cell-cycle control of gene expression in budding and fission yeast. *Annual Reviews in Genetics*, 39(1):69–94, 2005.

[49] Spellman PT, Sherlock G, Zhang MQ, Iyer VR, Andres K, Eisen MB, Brown PO, Botstein D, and Futcher B. Comprehensive identification of

cell cycle–regulated genes of the yeast *Saccharomyces cerevisiae* by microarray hybridization. *Molecular Biology of the Cell*, 9(1):3273–3297, 1998.

[50] Stark C, Breitkreutz BJ, Reguly T, Boucher L, Breitkreutz A, and Tyers M. BioGRID: A general repository for interaction datasets. *Nucleic Acids Research*, 34(1):D535–D539, 2006.

[51] Montefusco F, Cosentino C, Kim J, Amato F, and Bates DG. Reconstruction of partially-known biomolecular interaction networks from noisy data. In *Proceedings of the IFAC World Congress on Automatic Control*, Milan, Italy, 2011.

[52] Olson DL and Delen D. *Advanced Data Mining Techniques*. Berlin: Springer, 2008.

[53] Yu J, Smith VA, Wang PP, Hartemink AJ, and Jarvis ED. Advances to Bayesian network inference for generating causal networks from observational biological data. *Bioinformatics*, 20:3594–3603, 2004.

[54] Teixeira MC, et al. The YEASTRACT database: a tool for the analysis of transcription regulatory associations in *Saccharomyces cerevisiae*. *Nucleic Acids Research*, 34(1):D446–D451, 2006.

8

Stochastic effects in biological control systems

8.1 Introduction

At macroscopic scales, the processes of life appear highly deterministic. At molecular scales, however, biological processes are highly stochastic, due both to the cellular environment and the nature of information flows in biological networks. The effects of these stochastic variations or noise are neglected in deterministic chemical rate equations (and their corresponding differential equation models) and this is reasonable, because in most cases such effects disappear when they are averaged over large numbers of molecules and chemical reactions.

In the case of biological processes involving molecular species at very low copy numbers, however, potentially significant stochastic effects may arise due to the random variations in numbers of molecules present in different cells at different times, [1]. A striking example is the process of transcription, where only one or two copies of a particular DNA regulatory site may be present in each cell. Clearly, in this case, the implicit assumption that the reactants vary both continuously and differentiably, which underlies the formulation of deterministic models, does not hold. Indeed, recent research has shown how deterministic models of a genetic network may not correctly represent the evolution of the mean of the corresponding (actual) stochastic system, [2]. Other research has revealed how stochastic noise can also play an important role in the dynamics of other types of cellular networks. In developmental processes, for example, it is often highly desirable to control or buffer stochastic fluctuations, whereas in other situations noise allows organisms to generate non-genetic phenotypic variability which may confer robustness to changes in environmental conditions. Stochastic noise has been shown to have the potential to cause qualitative changes in the dynamics of some systems, for example causing random switching between different equilibria in systems exhibiting bistability, or inducing oscillations in otherwise stable systems, [3, 4].

The theoretical machinery required to rigorously analyse stochastic biomolecular networks is much less well developed than in the case of deterministic systems. For this reason, the main focus of this book has been on deterministic ODE-based models, for which many powerful analysis tools exist in systems and control theory. Some recent research has, however, led to the development of promising approaches for characterising the effects of stochastic noise on important system properties such as stability and robustness, and these are described later in this chapter. In many cases, however, the analysis of stochastic effects in biological control systems is still reliant on computer simulation, and so, following the treatment of [2], we begin with a brief summary of the basic modelling and simulation tools which are available for this purpose.

8.2   Stochastic modelling and simulation

Consider a system of molecules comprised of N chemical species $(S_1, ..., S_N)$ interacting via M chemical reaction channels $(R_1, ..., R_M)$. The system is assumed to be spatially homogenous (well-stirred), operating in a constant volume Ω and to be in thermal (but not chemical) equilibrium at some constant temperature. Denote by $X_i(t)$ the number of molecules of species S_i present in the system at time t. For some initial condition $X(t_0) = x_0$, we want to compute the evolution of the state vector $X(t)$ over time.

We assume that each reaction channel R_j describes a distinct physical event which happens essentially instantaneously and can be characterised mathematically by the quantities v_j and a_j. $v_j = (v_{1j}, ..., v_{Nj})$ is a state-change vector, and v_{ij} is defined as the change in the population of species S_i caused by the reaction R_j. Thus, the reaction R_j will cause an instantaneous change in the state of the system from some state $X(t) = x$ to state $x + v_j$, and the array (v_{ij}) is the system's stoichiometric matrix. a_j is called the propensity function, which is defined so that for a system at state $X(t) = x$, $a_j(x)dt$ is the probability that one R_j reaction will occur somewhere in the volume Ω in the next infinitesimal time interval $[t, t + dt)$. If R_j is the monomolecular reaction $S_i \rightarrow$ products, then quantum mechanics implies the existence of some constant c_j such that $a_j(x) = c_j x_i$. If R_j is the bimolecular reaction $S_i + S_{i'} \rightarrow$ products, then the underlying physics gives a different constant c_j, and a propensity function $a_j(x)$ of the form $c_j x_i x_{i'}$ if $i \neq i'$, or $c_j \frac{1}{2} x_i(x_i - 1)$ if $i = i'$, [5, 6].

Since in this framework the underlying bimolecular reactions are stochastic, the precise positions and velocities of all the molecules in the system are not known. Thus, it is only possible to compute the probability that an S_i molecule and an $S_{i'}$ molecule will collide in the next dt, and the probability that such a collision will result in an R_j reaction. For a monomolecular reaction, c_j is equal to the reaction rate constant k_j of conventional deterministic chemical kinetics, while for a bimolecular reaction c_j is equal to k_j/Ω if the reactants are different species, or $2k_j/\Omega$ if they are the same, [6, 7, 8].

Now, we want to compute the probability that $X(t)$ is equal to some value

x, given that $X(t_0)$ is equal to some value x_0, i.e. $P(x, t \mid x_0, t_0)$. A time-evolution equation for this probability is given by

$$P(x, t + dt \mid x_0, t_0) = P(x, t, \mid x_0, t_0) \times \left[1 - \sum_{j=1}^{M} a_j(x) dt \right] +$$

$$\sum_{j=1}^{M} P(x - v_j, t \mid x_0, t_0) \times a_j(x - v_j) dt$$

The first term on the right-hand side of the above equation is the probability that the system is already in state x at time t and no reaction of any kind occurs in the time interval $[t, t + dt)$. The second term is the probability that the system is one R_j reaction away from state x at time t and that one R_j reaction occurs in the interval $[t, t + dt)$. Note that dt is assumed to be so small that no more than one reaction of any kind can occur in the interval $[t, t + dt)$. Now, we subtract $P(x, t \mid x_0, t_0)$ from both sides of the above equation, divide through by dt, and take the limit as $dt \to 0$ to obtain [5, 7]

$$\frac{\delta P(x, t \mid x_0, t_0)}{\delta t} = \sum_{j=1}^{M} [a_j(x - v_j) P(x - v_j, t \mid x_0, t_0) - a_j(x) P(x, t \mid x_0, t_0)]$$

The above equation is known as the *chemical master equation* (CME), and it completely determines the function $P(x, t \mid x_0, t_0)$. Unfortunately, however, the CME consists of almost as many coupled ordinary differential equations as there are combinations of molecules that can exist in the system — it can only be solved analytically in the case of a few very simple systems, and even numerical solutions are usually prohibitively expensive computationally. A solution to this problem is provided by the *stochastic simulation algorithm* (SSA), which works by constructing numerical realisations of $X(t)$, i.e. simulated trajectories of $X(t)$ over time — when averaged over many realisations, the resulting trajectories represent good approximations to exact numerical solutions of the CME. The basic idea behind the SSA is to generate a new function, $p(\tau, j \mid x, t)$, [9], such that $p(\tau, j \mid x, t) d\tau$ is the probability, given $X(t) = x$, that the next reaction in the system will be R_j and that this reaction will occur in the infinitesimal time interval $[t + \tau, t + \tau + d\tau)$. This function is thus the joint probability density function of the two random variables τ (the time to the next reaction) and j (the index of the next reaction). An analytical expression for $p(\tau, j \mid x, t) d\tau$ can be derived as follows. First note that if $P_0(\tau \mid x, t)$ is the probability, given $X(t) = x$, that no reaction occurs in the time interval $[t, t + \tau)$ then we have that

$$p(\tau, j \mid x, t) d\tau = P_0(\tau \mid x, t) \times a_j(x) d\tau$$

$$P_0(\tau + d\tau \mid x, t) = P_0(\tau \mid x, t) \times \left[1 - \sum_{k=1}^{M} a_k(x) d\tau \right]$$

Rearranging the last equation and taking the limit as $d\tau \to 0$ gives a differential equation whose solution is easily found to be $P_0(\tau \mid x,t) = \exp(-a_0(x)\tau)$, where $a_0(x) \equiv \sum_{k=1}^{M} a_k(x)$. Inserting this into the previous equation gives

$$p(\tau, j \mid x, t) = a_j(x)\exp(-a_0(x)\tau)$$

Note that the above equation implies that the joint density function of τ and j can be written as the product of the τ-density function, $a_0(x)\exp(-a_0(x)\tau)$, and the j-density function, $a_j(x)/a_0(x)$. Using Monte Carlo theory, [9], random samples can be drawn from these two density functions as follows: generate two random numbers r_1 and r_2 from the uniform distribution in the unit interval and then select τ and j according to

$$\tau = \frac{1}{a_0(x)}\ln\frac{1}{r_1} \tag{8.1}$$

$$\sum_{k=1}^{j-1} a_k(x) \le r_2 a_0(x) < \sum_{k=1}^{j} a_k(x) \tag{8.2}$$

The SSA is then given as follows:

1. Initialise the time $t = t_0$ and the system's state $x = x_0$.
2. Evaluate all the $a_j(x)$ and their sum $a_0(x)$ with the system in state x at time t.
3. Generate values for τ and j according to Eqs. 8.1 and 8.2.
4. Simulate the next reaction by replacing t with $t + \tau$ and x with $x + v_j$.
5. Record the new values of (x,t) and return to Step 2, or else end the simulation.

The $X(t)$ trajectory that is produced by the SSA can be interpreted as a stochastic version of the trajectory that would be found by solving the standard reaction rate equation from deterministic chemical kinetics. Note also that the exact value of τ used at each time step in the SSA is different — in contrast to the time step used in most numerical solvers for deterministic simulations, τ is not a finite approximation to some infinitesimal dt. Although the SSA is very straightforward to implement, it is often very slow, due primarily to the factor $1/a_0 x$ in Eq. 8.1, which will be very small if the population of any reactant species is sufficiently large, as is often the case in practice.

Much research in recent years has been devoted to attempts to find more computationally efficient methods for the simulation of stochastic models. Several variations to the above method for implementing the SSA have been developed, some of which are more efficient than others, [10, 11]. Inevitably, however, any procedure that simulates every reaction event one at a time will be highly computationally intensive. This has prompted several researchers to develop more approximate but faster approaches. One approximate accelerated simulation strategy is tau-leaping, [12], which advances the system by a pre-selected time τ which encompasses more than one reaction event, but is still sufficiently small that no propensity function changes its value by a significant amount. τ-leaping has been shown to allow much faster simulation of some systems, [12, 13], but it can also lead to erroneous results if the chosen leaps are too large, [14]. In addition, large leaps cannot be taken in the case of "stiff" systems with widely varying time scales (which are very common in cellular systems) since the maximum allowable leap is limited by the time scale of the fastest mode.

Other related approaches are the Langevin leaping formula, [15], hybrid methods which combine deterministic simulation of fast reactions involving large populations with the use of the SSA to simulate slow reactions with small populations, [16], and finite state projection algorithms [17, 18].

8.3   A framework for analysing the effect of stochastic noise on stability

The development of efficient methods for the simulation of stochastic network models is clearly an important research direction in systems biology. As in the case of deterministic systems, however, analytical tools will also be required in order to obtain a detailed understanding of the design principles of such systems. Such tools will be of even more importance for the design of synthetic circuits, or of therapeutics aimed at altering existing networks, since in these cases a systematic mapping of the parameter space is required. For these purposes, it is likely that stochastic simulation will be prohibitively time consuming no matter what improvements in efficiency are produced, especially for large-scale networks. In light of this, it is perhaps surprising that the problem of characterising the effects of stochastic noise on system properties has to date received relatively little attention from systems and control theorists.

One approach which does appear to have significant potential is based on the idea of incorporating approximate models of noise into deterministic modelling frameworks using the so-called linear noise approximation, [19]. In particular, a recent extension of this approach, termed the "effective stability approximation" method, [20], represents a potentially powerful framework for the analysis of the effect of intrinsic noise on the stability of biological systems.

8.3.1   The effective stability approximation

The basic idea of this approach is as follows. Consider a system of biomolecular interactions represented by the nonlinear differential equation

$$\frac{dx(t)}{dt} = f[x(t)] \tag{8.3}$$

where $x \in \mathbb{R}^n$, $f[x(t)]$ satisfies the standard conditions for the existence and uniqueness of the solution of the differential equation, $\mathbb{R}$ is the real number field and n is a positive integer. Linear stability analysis of such equations is performed around the equilibrium point, x_s, which satisfies $f(x_s) = 0$, as follows:

$$\frac{d\Delta x(t)}{dt} = \left. \frac{\partial f(x)}{\partial x} \right|_{x=x_s} \Delta x(t) \equiv \Gamma \Delta x(t) \tag{8.4}$$

where we assume that all real parts of the eigenvalues of Γ are strictly less than zero, hence the system is Hurwitz stable. Now, introduce a small perturbation which is added to $\Delta x(t)$, to represent some level of stochastic noise $\Omega \alpha(t)$, where Ω in the set of positive real numbers, $\mathbb{R}^+$, is in general inversely proportional to the square root of the cell volume, V_{cell}, i.e. $\Omega \approx 1/\sqrt{V_{\text{cell}}}$, and $\alpha(t)$ in $\mathbb{R}^n$ is the stochastic noise whose mean value is zero. Then, the above perturbation including the stochastic fluctuation can be approximated as follows:

$$\frac{d\delta x(t)}{dt} \approx \Gamma \delta x(t) + \Omega \left. \frac{\partial}{\partial \Omega} \left[\left. \frac{\partial f(x)}{\partial x} \right|_{x=x_s+\Omega\alpha(t)} \right] \right|_{\Omega=0} \delta x(t)$$

$$\equiv \Gamma \delta x(t) + \Omega J[\alpha(t)]\delta x(t) \tag{8.5}$$

We are interested in the mean trajectory of $\delta x(t)$, which is given by:

$$\frac{d\mathbf{E}\left[\delta x(t)\right]}{dt} \approx \Gamma \mathbf{E}\left[\delta x(t)\right] + \Omega \mathbf{E}\left\{ J[\alpha(t)]\delta x(t) \right\} \tag{8.6}$$

where $\mathbf{E}(\cdot)$ is the expectation. The following Bourret's approximation can be derived by assuming that $\alpha(t)$ varies much faster than $e^{-\Gamma t}\delta x(t)$ and neglecting the terms in Ω higher than second order, [21]:

$$\frac{d\mathbf{E}\left[\delta x(t)\right]}{dt} \approx \Gamma \mathbf{E}\left[\delta x(t)\right] + \Omega^2 \int_0^t \mathbf{E}\left[J_c(t-\tau)\right] \mathbf{E}\left[\delta x(\tau)\right] d\tau \tag{8.7}$$

where $J_c(t-\tau) = J[\alpha(t)]e^{\Gamma(t-\tau)}J[\alpha(\tau)]$. Note that each term of $J_c(t-\tau)$ is a linear combination of $\alpha_i(t)\alpha_j(\tau)$, $\alpha_i(t)$ is the i-th element of $\alpha(t)$ and the covariance of $\alpha(t)$ is derived from linearised Fokker–Plank equations as follows:

$$\mathbf{E}\left[\alpha(t)\alpha^T(\tau)\right] = e^{\Gamma(t-\tau)}\Xi, \tag{8.8}$$

where Ξ is given by the solution of the Lyapunov equation:

$$\Gamma \Xi + \Xi \Gamma^T + D = 0, \tag{8.9}$$

$D = S\text{diag}[v]S^T$, $f(x) = Sv$ and S is the stoichiometry matrix for the network, [22]. Then, $J_c(t - \tau)$ is a function of the time difference only. Since the integral in the right-hand side of Eq. (8.7) is a convolution integral, the Laplace transform of both sides is given by [20]

$$\delta X(s) = \left[sI - \Gamma - \Omega^2 \hat{J}_c(s)\right]^{-1} \delta X(0) \qquad (8.10)$$

where I is the identity matrix and $\hat{J}_c(s)$ is the Laplace transform of $\mathbf{E}\left[J_c(t)\right]$. The effect of the stochastic noise on the stability of the system can now be analysed using this equation; however, notice that in the calculation of $J_c(t)$ the symbol t is involved in calculating the matrix exponential, $e^{\Gamma t}$. The calculation of this matrix exponential will thus be extremely computationally expensive, [23], and in practice restricts the method as formulated in [20] to the analysis of very small-scale circuits, of the order of two or three states at most. To extend the applicability of the approach to larger size problems, a novel approximation for the dominant term in the stochastic perturbation to the ordinary differential equation model was developed in [24], as described below.

8.3.2 A computationally efficient approximation of the dominant stochastic perturbation

Recall that the original linearised differential equation is assumed to be stable, i.e. $e^{\Gamma t} \to 0$ as $t \to \infty$. For all Hurwitz stable Γ and any δ greater than zero, it is easy to show that there always exists a positive number, τ_c, such that

$$\left\| \mathbf{E} \left\{ J[\alpha(t)]e^{\Gamma t} J[\alpha(0)] \right\} \right\| < \delta, \qquad (8.11)$$

for all $t > \tau_c$. Then, the Bourret's representation may be approximated as follows:

$$\frac{d\mathbf{E}\left[\delta x(t)\right]}{dt} \approx \Gamma \mathbf{E}\left[\delta x(t)\right] + \Omega^2 \int_0^t \mathbf{E}\left[T_c(t)\right] \mathbf{E}\left[\delta x(t - \tau)\right] d\tau \qquad (8.12)$$

where

$$T_c(t) = \begin{cases} J_c(t), & \text{for } t \le \tau_c \\ 0, & \text{for } t > \tau_c \end{cases} \qquad (8.13)$$

and the approximation error is bounded by

$$(\text{approximation error}) \le \Omega^3 \left\| \int_{\tau_c}^t \mathbf{E}\left[\delta x(t - \tau)\right] d\tau \right\| \qquad (8.14)$$

This approximation does not introduce any significant additional error beyond the level of approximation that is imposed in the standard Bourret's representation. To see this, split the integral in Eq. (8.12) into two subintervals, i.e. $\tau \in [0, \tau_c)$ and $\tau \in [\tau_c, t)$. Setting δ equal to Ω, we have that the integral from $\tau = \tau_c$ to $\tau = t$ is bounded by

$$\Omega^2 \left\| \int_{\tau_c}^t \mathbf{E}\left[J_c(t)\right] \mathbf{E}\left[\delta x(t-\tau)\right] d\tau \right\|$$

$$= \Omega^2 \left\| \int_{\tau_c}^t \mathbf{E}\left\{J[\alpha(t)]e^{\Gamma t}J[\alpha(0)]\right\} \mathbf{E}\left[\delta x(t-\tau)\right] d\tau \right\|$$

$$\leq \Omega^2 \left\| \int_{\tau_c}^t \left\| \mathbf{E}\left\{J[\alpha(t)]e^{\Gamma t}J[\alpha(0)]\right\}\right\| \mathbf{E}\left[\delta x(t-\tau)\right] d\tau \right\|$$

$$= \Omega^3 \left\| \int_{\tau_c}^t \mathbf{E}\left[\delta x(t-\tau)\right] d\tau \right\| \tag{8.15}$$

Since the standard Bourret's representation ignores all terms higher than Ω^2, no significant additional error is introduced in the approximation. Note that since the local stability around the equilibrium point is checked by inspecting the eigenvalues of the perturbed equation, the norm of the perturbed state is assumed to be sufficiently small so that the last integration of the perturbed state in Eq. (8.15) from time τ_c to t remains smaller than $1/\Omega$. The stability of the stochastic network can thus be checked by analysing the following equation:

$$\delta X(s) = \left[sI - \Gamma - \Omega^2 \hat{T}_c(s)\right]^{-1} \delta X(0) \tag{8.16}$$

where

$$\hat{T}_c(s) = \int_0^\infty \mathbf{E}\left[T_c(t)\right] e^{-st} dt = \int_0^{\tau_c} \mathbf{E}\left[J_c(t)\right] e^{-st} dt \tag{8.17}$$

It is still difficult in general to obtain an exact closed form solution for this integral, but it can be approximated numerically by using the following result. The Laplace transform of $T_c(t)$ is given by

$$\hat{T}_c(s) = \sum_{k=1}^N F_k\left[k\Delta t, \Gamma, \Xi\right] \frac{e^{-s(k-1)\Delta t} - e^{-sk\Delta t}}{s} \tag{8.18}$$

where

$$F_k\left[k\Delta t, \Gamma, \Xi\right] = \mathbf{E}\left\{J[\alpha(k\Delta t)]e^{\Gamma k \Delta t}J[\alpha(0)]\right\} \tag{8.19}$$

$\Delta t = \tau_c/N$ and the error between $\hat{T}_c(s)$ and the Laplace transform of $J_c(t)$ can be made arbitrarily small for all $s = j\omega$, $\omega \in [0, \infty)$ by increasing N and τ_c while keeping Δt small. The matrix exponential $e^{\Gamma k \Delta t}$ is approximated by $[I + (\Delta t/r)\Gamma]^{rk}$ and $\mathbf{E}\left[\alpha(t)\alpha^T(0)\right]$ is approximated by $[I + (\Delta t/r)\Gamma]^{rk} \Xi$, where r is a positive real number greater than or equal to Δt. To see that the approximation error can be made arbitrarily small, note that to obtain an approximate integral, the interval from 0 to τ_c is divided into the sum of N subintervals, whose length equals $\Delta t = \tau_c/N$ such that

$$\mathbf{E}\left\{J[\alpha(t)]e^{\Gamma t}J[\alpha(0)]\right\} \approx \mathbf{E}\left\{J[\alpha(k\Delta t)]e^{\Gamma k \Delta t}J[\alpha(0)]\right\} \tag{8.20}$$

for a sufficiently large N, for all $t \in [(k-1)\Delta t, k\Delta t)$. Then

$$
\begin{aligned}
\hat{T}_c(s) &= \int_0^{T_c} \mathbf{E}\left[J_c(t)\right] e^{-st} dt \\
&= \sum_{k=1}^{N} \int_{(k-1)\Delta t}^{k\Delta t} \mathbf{E}\left\{J[\alpha(t)]e^{\Gamma t}J[\alpha(0)]\right\} e^{-st} dt \\
&\approx \sum_{k=1}^{N} \mathbf{E}\left\{J[\alpha(k\Delta t)]e^{\Gamma k\Delta t}J[\alpha(0)]\right\} \frac{e^{-s(k-1)\Delta t} - e^{-sk\Delta t}}{s}
\end{aligned}
\tag{8.21}
$$

where the matrix exponential for k is approximated as mentioned above. The approximation error for $\hat{T}_c(s)$ is bounded by

$$
\left\| \sum_{k=1}^{N} \int_{(k-1)\Delta t}^{k\Delta t} \left\{\mathbf{E}\left[J_c(k\Delta t)\right] - \mathbf{E}\left[J_c(t)\right]\right\} e^{-st} dt \right\|
$$

$$
\leq \Delta t^2 \sum_{k=1}^{N} \Delta J_k \leq \frac{T_c^2}{N} \Delta \bar{J}
\tag{8.22}
$$

where the first inequality is satisfied because the integral of e^{-st} for the given interval is bounded by Δt, ΔJ_k is the maximum of $\|\mathbf{E}\left[J_c(k\Delta t)\right] - \mathbf{E}\left[J_c(t)\right]\|$ for $t \in [(k-1)\Delta t, k\Delta t)$ and $\Delta \bar{J}$ is the maximum of ΔJ_k for $k \in [1, N]$. Thus, as N and r grow, $\Delta \bar{J}$ converges to zero and the approximation error approaches zero.

Hence, the stability of the stochastic network may be checked via the following characteristic equation:

$$
\left| sI - \Gamma - \Omega^2 \sum_{k=1}^{N} F_k\left[\Delta t, \Gamma, \Xi\right] \frac{e^{-s(k-1)\Delta t} - e^{-sk\Delta t}}{s} \right| = 0
\tag{8.23}
$$

where $|\cdot|$ is the determinant and Δt is a fixed positive real number.

8.3.3 Analysis using the Nyquist stability criterion

Before proceeding, we note several properties of the irrational term in Eq. (8.23), $(e^{-s(k-1)\Delta t} - e^{-sk\Delta t})/s$.

1. The irrational term is analytic over the whole complex plane.

2. The magnitude is bounded by Δt.

3. The irrational term is BIBO (bounded input bounded output) stable.

For stability analysis, we need to check the signs of the real parts of all roots of the characteristic equation and hence we need to obtain all roots of

Eq. (8.23). To avoid dealing with infinite polynomials, we first write Eq. (8.16) as follows:

$$\delta X(s) = [I - M(s)\Delta(s)]^{-1} M(s)\delta X(0) \qquad (8.24)$$

where $M(s) = [sI - \Gamma]^{-1}$ and $\Delta(s) = \Omega^2 \hat{T}_c(s)$. Note that since the irrational term is BIBO stable, $\Delta(s)$ does not have any pole in the right half of the complex plane. Also, since the irrational term is analytic on the whole complex plane, it does not affect the number of encirclements of the origin. Therefore, the following result is an immediate consequence of the application of the generalised Nyquist stability criterion:

Let τ_c be generated as described in the previous section, let N be a sufficiently large integer and let $\Delta t = \tau_c/N$. Then the deterministic differential equation, (8.4), is stable with respect to stochastic perturbation $\Delta(s) = \Omega^2 \hat{T}_c(s)$, where $\hat{T}_c(s)$ is defined in Eq. (8.17), if and only if

$$\left| I - M(j\omega)\Omega^2 \sum_{k=1}^{N} F_k\left[\Delta t, \Gamma, \Xi\right] \frac{e^{-j\omega(k-1)\Delta t} - e^{-j\omega\Delta t}}{j\omega} \right| \qquad (8.25)$$

does not encircle the origin for $\forall \omega \in (-\infty, \infty)$.

Checking the above necessary and sufficient condition for stability involves counting the number of encirclements of the origin made by the Nyquist plot, which can sometimes be cumbersome, and requires a certain number of frequency evaluations. The following sufficient conditions for stability, which are direct consequences of the Nyquist stability criterion and the triangle inequality, can be checked even more efficiently, at the expense of some possible conservatism.

Let the norm of $M(j\omega)$ be bounded by a positive real number, γ, for all $\omega \in [0, \infty)$ and let τ_c, N and $\Delta t = \tau_c/N$ be as above. The deterministic differential equation, (8.4), is stable with respect to the stochastic perturbation $\Delta(s) = \Omega^2 \hat{T}_c(s)$, where $\hat{T}_c(s)$ is defined in Eq. (8.17), if either of the following holds:

$$\|M(j\omega)\Delta(j\omega)\|$$
$$= \Omega^2 \left\| M(j\omega) \sum_{k=1}^{N} F_k\left[\Delta t, \Gamma, \Xi\right] \frac{e^{-j\omega(k-1)\Delta t} - e^{-j\omega\Delta t}}{j\omega} \right\|$$
$$\leq 1 \qquad (8.26)$$

or

$$\Omega^2 \gamma \Delta t \sum_{k=1}^{N} \left\| F_k\left[\Delta t, \Gamma, \Xi\right] \right\| \leq 1 \qquad (8.27)$$

The results presented above provide a striking example of how classical analysis methods from control engineering can be adapted to provide powerful tools for the analysis of stochastic biological systems. Other recent research has fused control theory with information theory to generate important new results characterising the "limits of performance" of cellular systems when it comes to dealing with stochastic noise, [25]. Clearly in this, and many other fields of biological research, we are only just beginning to exploit the huge potential of ideas and methods from systems and control engineering.

8.4   Case Study XIII: Stochastic effects on the stability of cAMP oscillations in aggregating *Dictyostelium* cells

We consider the same deterministic model for cAMP oscillations in aggregating *Dictyostelium* cells which was used in Case Study II, [26]

$$d[\text{ACA}]/dt = k_1[\text{CAR1}] - k_2[\text{ACA}][\text{PKA}]$$
$$d[\text{PKA}]/dt = k_3[\text{cAMPi}] - k_4[\text{PKA}]$$
$$d[\text{ERK2}]/dt = k_5[\text{CAR1}] - k_6[\text{PKA}][\text{ERK2}]$$
$$d[\text{RegA}]/dt = k_7 - k_8[\text{ERK2}][\text{RegA}] \qquad (8.28)$$
$$d[\text{cAMPi}]/dt = k_9[\text{ACA}] - k_{10}[\text{RegA}][\text{cAMPi}]$$
$$d[\text{cAMPe}]/dt = k_{11}[\text{ACA}] - k_{12}[\text{cAMPe}]$$
$$d[\text{CAR1}]/dt = k_{13}[\text{cAMPe}] - k_{14}[\text{CAR1}]$$

where ACA is adenylyl cyclase, PKA is the protein kinase, ERK2 is the mitogen activated protein kinase, RegA is the cAMP phosphodiesterase, cAMPi and cAMPe are the internal and the external cAMP concentrations, respectively, and CAR1 is the cell receptor. Uncertainty in each kinetic parameter in the model is represented as $k_i = \bar{k}_i (1 + p_\delta \delta_i/100)$ for $i = 1, 2, \ldots, 13, 14$. $\bar{k}_i$ is the nominal value of each k_i, which are given by [27, 28]: $\bar{k}_1 = 2.0$ min^{-1}, $\bar{k}_2 = 0.9$ μM^{-1}min^{-1}, $\bar{k}_3 = 2.5$ min^{-1}, $\bar{k}_4 = 1.5$ min^{-1}, $\bar{k}_5 = 0.6$ min^{-1}, $\bar{k}_6 = 0.8$ μM^{-1}min^{-1}, $\bar{k}_7 = 1.0$ μM min^{-1}, $\bar{k}_8 = 1.3$ μM^{-1}min^{-1}, $\bar{k}_9 = 0.3$ min^{-1}, $\bar{k}_{10} = 0.8$ μM^{-1}min^{-1}, $\bar{k}_{11} = 0.7$ min^{-1}, $\bar{k}_{12} = 4.9$ min^{-1}, $\bar{k}_{13} = 23.0$ min^{-1} and $\bar{k}_{14} = 4.5$ min^{-1}, while δ_i represents uncertainty in the kinetic parameters. In [29], the worst-case direction for perturbations in the parameter space which destroy the stable limit cycle was identified as $\delta_1 = -1$, $\delta_2 = -1$, $\delta_3 = 1$, $\delta_4 = 1$, $\delta_5 = -1$, $\delta_6 = 1$, $\delta_7 = 1$, $\delta_8 = -1$, $\delta_9 = 1$, $\delta_{10} = 1$, $\delta_{11} = -1$, $\delta_{12} = 1$, $\delta_{13} = -1$ and $\delta_{14} = 1$. p_δ represents the magnitude of the parameter-space perturbation in percent.

For p_δ equal to zero, the above set of differential equations exhibits a stable limit cycle. However, for values of p_δ greater than 0.6, the equilibrium point becomes stable and the limit cycle disappears. Here, we are going to study whether this is also true for the corresponding stochastic model.

To transform the above ordinary differential equations into the corresponding stochastic model, the following fourteen chemical reactions are deduced from the ODE model, [1]:

$$\text{CAR1} \xrightarrow{k_1} \text{ACA} + \text{CAR1},$$

$$\text{ACA} + \text{PKA} \xrightarrow{k_2/n_A/\mathcal{V}/10^{-6}} \text{PKA},$$

$$\text{cAMPi} \xrightarrow{k_3} \text{PKA} + \text{cAMPi},$$

$$\text{PKA} \xrightarrow{k_4} \emptyset,$$

$$\text{CAR1} \xrightarrow{k_5} \text{ERK2} + \text{CAR1},$$

$$\text{PKA} + \text{ERK2} \xrightarrow{k_6/n_A/\mathcal{V}/10^{-6}} \text{PKA},$$

$$\emptyset \xrightarrow{k_7 \times n_A \times \mathcal{V} \times 10^{-6}} \text{RegA}, \tag{8.29}$$

$$\text{ERK2} + \text{RegA} \xrightarrow{k_8/n_A/\mathcal{V}/10^{-6}} \text{ERK2},$$

$$\text{ACA} \xrightarrow{k_9} \text{cAMPi} + \text{ACA},$$

$$\text{RegA} + \text{cAMPi} \xrightarrow{k_{10}/n_A/\mathcal{V}/10^{-6}} \text{RegA},$$

$$\text{ACA} \xrightarrow{k_{11}} \text{cAMPe} + \text{ACA},$$

$$\text{cAMPe} \xrightarrow{k_{12}} \emptyset,$$

$$\text{cAMPe} \xrightarrow{k_{13}} \text{CAR1} + \text{cAMPe},$$

$$\text{CAR1} \xrightarrow{k_{14}} \emptyset,$$

where $\emptyset$ represents some relatively abundant source of molecules or a non-interacting product, n_A is Avogadro's number, 6.023×10^{23}, $\mathcal{V}$ is the average volume of a *Dictyostelium* cell, $0.565 \times 10^{-12} l$, [30], and 10^{-6} is a multiplication factor due to the unit μM. The probability of each reaction occurring is defined by the rate of each reaction. For example, the probabilities during a small length of time, dt, that the first and the second reactions occur are given by $k_1 \times \text{CAR1}$ and $k_2/n_A/\mathcal{V}/10^{-6} \times \text{ACA} \times \text{PKA}$, respectively. The probabilities for all the other reactions are defined similarly. To conduct stochastic simulations of this system, the chemical master equation was obtained and solved approximately using standard software implementations of the SSA, [8].

For the Bourret's approximation, the system volume, V_{cell}, has the following relation to the density and the number of molecules,

$$V_{\text{cell}} = x \frac{(\text{\# of molecules})}{\mu\text{M}} = 1\mu\text{M} \times \mathcal{V}$$

$$= \frac{10^{-6} \times 6.023 \times 10^{23}}{\text{liter}} \times \mathcal{V} = 3.403 \times 10^5 \tag{8.30}$$

For this problem, since the state dimension is seven, calculating the matrix exponential symbolically as required in the original formulation of the effective

stability approximation in [20] is not computationally feasible. Hence, the
new approximation proposed in [24] has to be used. With N fixed at 200, τ_c
is chosen such that $\tau_c = \ln 0.01/\max_{i=1,2,\ldots 7} \Re(\lambda_i)$, where $\Re(\lambda_i)$ is the real
part of the eigenvalues of Γ and r, the number of intervals to approximate the
exponential function, is chosen equal to 1000. The Nyquist plot for this system
is shown for $p_\delta = 0.6$ in Fig. 8.1.   As shown in Fig. 8.2, the deterministic

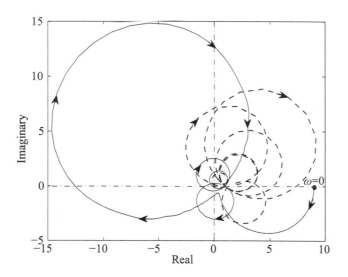

FIGURE 8.1: Nyquist plot: $p_\delta = 0.6$.

model with this set of parameter values converges to a steady-state and ceases
to oscillate. However, since the Nyquist plot given in Fig. 8.1 has more than
one encirclement of the origin, the system cannot converge to a steady-state if
the stochastic effect is taken into account. The stochastic simulations shown
in Fig. 8.2 using Gillespie's direct method confirm this result, i.e the model
including stochastic noise continues to oscillate. We note that this result is
of independent biological interest, since it represents an example of stochastic
noise changing the qualitative behaviour of a network model even at very
high molecular concentrations. Here, however, we are primarily interested in
the computational complexity of the stability calculation. It takes about 54
hours to perform the stochastic simulation; however, the proposed analytical
method for determining the stability of the stochastic model gives the answer
in less than 1 hour.

When the magnitude of the perturbation in the model's parameters p_δ
is increased to 1.5 and 2.0, the analysis results are shown in Figs. 8.3 and

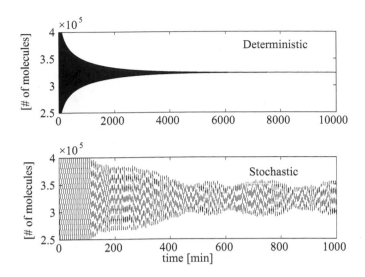

FIGURE 8.2: The internal cAMP time history: $p_\delta = 0.6$.

8.4. In both cases, the first sufficient condition for stability, Eq. (8.26), is now satisfied. For $p_\delta = 2.0$ the second sufficient condition is also satisfied as the left hand side of Eq. (8.27) is approximately equal to 0.3. We can thus conclude that the stochastic model will be stable (i.e. will not oscillate) without even checking the Nyquist plot. The stochastic time histories shown for both cases, of course, do not converge exactly to steady-states in a deterministic sense because of the existence of noise. However, the oscillation amplitudes are almost negligible compared to the case of $p_\delta = 0.6$ and therefore we can conclude that these two cases are not oscillating. The calculation time for the stochastic simulations for both cases is about 15 hours while for the Nyquist analysis the computations take less than 15 minutes (on a 3.06 GHz Pentium IV machine with 1GB of RAM).

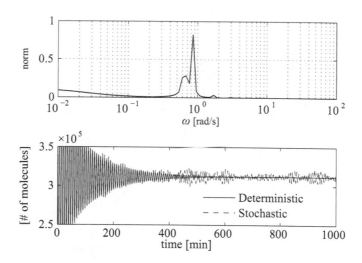

FIGURE 8.3: The sufficient condition and the internal cAMP time histories of the deterministic and the stochastic simulations for $p_\delta = 1.5$.

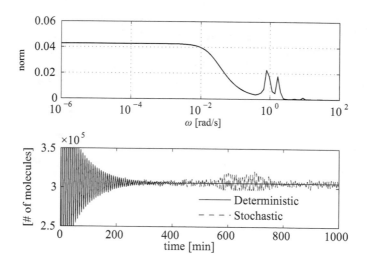

FIGURE 8.4: The sufficient condition and the internal cAMP time histories of the deterministic and the stochastic simulations for $p_\delta = 2.0$.

8.5 Case Study XIV: Stochastic effects on the robustness of cAMP oscillations in aggregating *Dictyostelium* cells

In the previous case study, we observed that stochastic noise could act to change the stability properties of a biochemical network underlying the generation of stable cAMP oscillations in aggregating *Dictyostelium* cells. For particular sets of parameter values, we were able to establish that the stochastic model would oscillate, while the corresponding deterministic model would not. By itself, this result does not say anything conclusive about the effect of noise on the robustness properties of the network, since it could simply be that different sets of parameter values are required to make the deterministic and stochastic models oscillate. In this case study, which is based on the results in [31], we consider the same network and systematically compare the robustness properties of the two models.

Since there are currently no analytical tools available with which to quantify the effect of noise on the robustness properties of biochemical networks, we resort to a simulation-based robustness analysis technique which is widely used in control engineering, Monte Carlo simulation. We generate 100 random samples of the kinetic constants, the cell volume and initial conditions from uniform distributions around the nominal values for several different uncertainty ranges. The kinetic constants are sampled uniformly from the following:

$$k_j^i = \bar{k}_j \left(1 + p_\delta \delta_j^i\right) \tag{8.31}$$

for $i = 1, 2, \ldots, n_c - 1, n_c$ and $j = 1, 2, \ldots, 13, 14$, where $\bar{k}_j^i$ is the nominal value of k_j, p_δ is the level of perturbation, i.e. 0.05, 0.1, or 0.2, δ_j^i is a uniformly distributed random number between -1 and $+1$ and n_c is the number of cells. The initial condition for internal cAMP is randomly sampled from the following:

$$\text{cAMPi}^i = \overline{\text{cAMPi}}^i \left(1 + p_\delta \delta_{\text{cAMPi}}^i\right) \tag{8.32}$$

for $i = 1, 2, \ldots, n_c - 1, n_c$, where $\overline{\text{cAMPi}}$ is the nominal initial value of cAMPi for the i-th cell and δ_{cAMPi}^i is a uniformly distributed random number between -1 and $+1$. The sampling for the other molecules is defined similarly. The nominal initial value for each molecule is given by [26] as: $\overline{\text{ACA}} = 7290, \overline{\text{PKA}} = 7100, \overline{\text{ERK2}} = 2500, \overline{\text{RegA}} = 3000, \overline{\text{cAMPi}} = 4110, \overline{\text{cAMPe}} = 1100$ and $\overline{\text{CAR1}} = 5960$. Similarly, the cell volume is perturbed as follows:

$$V = \bar{V} \left(1 + p_\delta \delta_V^i\right) \tag{8.33}$$

where $\bar{V} = 3.672 \times 10^{14}$ l and δ_V^i is a uniformly distributed random number between -1 and 1.

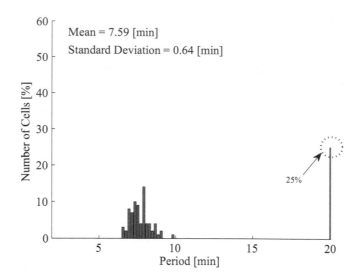

FIGURE 8.5: Robustness analysis of the period of the internal cAMP oscillations with respect to perturbations of 20% in the model parameters and initial conditions: deterministic model.

The simulations for the deterministic model and the stochastic model are performed using the Runge-Kutta 5th order adaptive algorithm and the τ-leap complex algorithm, [12], with maximum allowed relative errors of 1×10^{-4} and 5×10^{-5}, respectively, which are implemented in the software Dizzy, version 1.11.4, [32]. From the simulations, the time series of the internal cAMP concentration is obtained with a sampling interval of 0.01 min from 0 to 200 min. Taking the Fourier transform using the fast Fourier transform command in MATLAB®, [33], the maximum peak amplitude is checked and the period is calculated from the corresponding peak frequency. If the peak amplitude is less than 10% of the bias signal amplitude, the signal is considered to be non-oscillatory.

The robustness of the period of the oscillations generated by the deterministic and stochastic models was compared for several different levels of uncertainty in the kinetic parameters. In each case the level of robustness observed was significantly higher for the stochastic model. Sample results are shown in Figs. 8.5 and 8.6 for a 20% level of uncertainty in the kinetic parameters. In the figures, the peak at the 20 minute period denotes the total number of cases where the trajectories converged to some steady-state value, i.e. failed to oscillate. Similar improvements in the robustness of the amplitude distributions were found in all cases, [31].

One important mechanism which is missing in the model of [26] is the com-

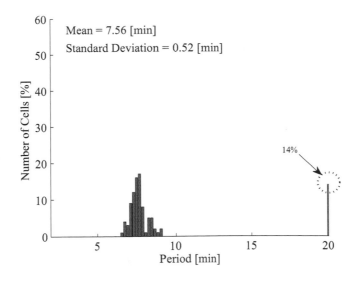

FIGURE 8.6: Robustness analysis of the period of the internal cAMP oscillations with respect to perturbations of 20% in the model parameters and initial conditions: stochastic model.

munication between neighbouring *Dictyostelium* cells through the diffusion of extracellular cAMP. During aggregation, *Dictyostelium* cells not only emit cAMP through the cell wall but also respond to changes in the concentration of the external signal which result from the diffusion of cAMP from large numbers of neighbouring cells. In [34], it was clarified how cAMP diffusion between neighbouring cells is crucial in achieving the synchronisation of the oscillations required to allow aggregation. Interestingly, similar synchronisation mechanisms have been observed in the context of circadian rhythms — the consequences and implications of such mechanisms are discussed in [35].

In order to investigate the effect of synchronisation on the robustness of cAMP oscillations in *Dictyostelium*, the stochastic version of the model of [26] must be modified to capture the interactions between cells. To consider synchronisation between multiple cells, the set of fourteen chemical reactions in the model is extended under the assumption that the distance between cells is small enough that diffusion is fast and uniform. In this case, the reactions for each individual cell just need to be augmented with one reaction that includes the effect of external cAMP emitted by all the other cells. Since the external cAMP diffuses fast and uniformly, the reaction involving k_{13} is modified as follows:

$$\text{cAMPe}/n_c \xrightarrow{k_{13_i}} \text{CAR1}_i + \text{cAMPe}/n_c \qquad (8.34)$$

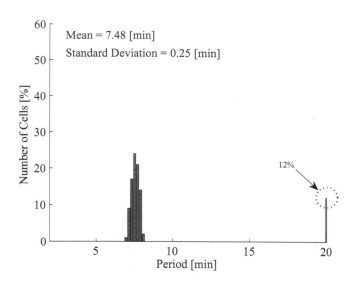

FIGURE 8.7: Robustness analysis of the period of the internal cAMP oscillations with respect to perturbations of 20% in the model parameters and initial
conditions: extended stochastic model with five synchronised cells.

for $i = 1, 2, \ldots, n_c - 1, n_c$, where cAMPe is the total number of external
cAMP molecules emitted by all the interacting cells, n_c is the total number
of cells, k_{13}^i is the i-th cell's kinetic constant for binding cAMP to CAR1 and
CAR1$_i$ is the i-th cell's CAR1 number.   Sample robustness analysis results
for the extended stochastic model in the case of five and ten interacting cells
with a 20% level of uncertainty are shown in Figs. 8.7 and 8.8. For all levels of
uncertainty, the resulting variation in the period of the oscillations reduces as
the number of synchronised cells in the extended model increases. Similar results were found for variations in the amplitude of the oscillations. Because of
the computational complexity of stochastic simulation, the maximum number of interacting cells that could be considered in the above analysis was
limited to ten. In reality, some 10^5 *Dictyostelium* cells form aggregates leading to slug formation, and each cell potentially interacts with far more than
ten other cells. The analysis of the stochastic model presented here suggests
how either direct or indirect interactions will lead to even stronger robustness of the cAMP oscillations as well as entrapment and synchronisation of
additional cells. The dependence of the dynamics of the cAMP oscillations
on the strength of synchronisation between the individual cells, as well as on
the level of cell-to-cell variation, may be critical mechanisms for developing
morphogenetic shapes in *Dictyostelium* development. In [36], for example, it
was shown experimentally that cell-to-cell variations desynchronise the devel-

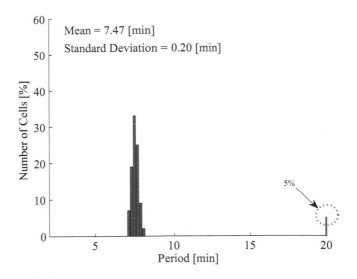

FIGURE 8.8: Robustness analysis of the period of the internal cAMP oscillations with respect to perturbations of 20% in the model parameters and initial conditions: extended stochastic model with ten synchronised cells.

opmental path and it was argued that they represent the key factor in the development of spiral patterns of cAMP waves during aggregation.

The results of this case study make some interesting contributions to the "stochastic versus deterministic" modelling debate in systems biology. Generally speaking, the arguments in favour of employing stochastic modelling frameworks have focused on the case of systems involving small numbers of molecules, where large variabilities in molecular populations favour a stochastic representation. Of course, this immediately raises the question of what exactly is meant by "small numbers" — see [37] for an interesting discussion of this issue. Here, we have analysed a system in which molecular numbers are very large, but the choice of a deterministic or stochastic representation still makes a significant difference to the robustness properties of the network model. The implications are clear — when using robustness analysis to check the validity of models for oscillating biomolecular networks, stochastic models should be used. The reason for this is that intracellular stochastic noise can constitute an important *source* of robustness for oscillatory biomolecular networks, and therefore must be taken into account when analysing the robustness of any proposed model for such a system. Finally, we showed how biological systems which are composed of networks of individual stochastic oscillators can use diffusion and synchronisation to produce wave patterns which are highly robust to variations among the components of the network.

References

[1] Wilkinson DJ. *Stochastic Modelling for Systems Biology.* Boca Raton: CRC Press, Taylor & Francis, 2006.

[2] El Samad H, Khammash M, Petzold L, and Gillespie D. Stochastic modeling of gene regulatory networks. *International Journal of Robust and Nonlinear Control*, 15:691–711, 2005.

[3] Kaern M, Elston TC, Blake WJ, and Collins JJ. Stochasticity in gene expression: From theories to phenotypes. *Nature Reviews Genetics*, 6(6):451–464, 2005.

[4] Raser JM and O'Shea EK. Noise in gene expression: origins, consequences, and control. *Science*, 309(5743):2010–2013, 2005.

[5] Gillespie D. A rigorous derivation of the chemical master equation. *Physica A*, 188:404–425, 1992.

[6] Gillespie D. *Markov Processes: An Introduction for Physical Scientists.* Boston: Academic Press, 1992.

[7] McQuarrie D. Stochastic approach to chemical kinetics. *Journal of Applied Probability*, 4:413–478, 1967.

[8] Gillespie D. Exact stochastic simulation of coupled chemical reactions. *Journal of Physical Chemistry*, 81:2340–2361, 1977.

[9] Gillespie D. A general method for numerically simulating the stochastic time evolution of coupled chemical reactions. *Journal of Computational Physics*, 22(4):403–434, 1976.

[10] Gibson M and Bruck J. Efficient exact stochastic simulation of chemical systems with many species and many channels. *Journal of Physical Chemistry*, 104:1876–1889, 2000.

[11] Cao Y, Li H, and Petzold L. Efficient formulation of the stochastic simulation algorithm for chemically reacting systems. *Journal of Chemical Physics*, 121(9):4059–4067, 2004.

[12] Gillespie D. Approximate accelerated stochastic simulation of chemically reacting systems. *Journal of Chemical Physics.* 115:1716–1733, 2001.

[13] Gillespie D and Petzold L. Improved leap-size selection for accelerated stochastic simulation. *Journal of Chemical Physics*, 119:8229–8234, 2003.

[14] Cao Y, Gillespie DT, and Petzold LR. Avoiding negative populations in explicit Poisson tau-leaping. *Journal of Chemical Physics*, 123:054104, 2005.

[15] Gillespie D. The chemical Langevin equation. *Journal of Chemical Physics*, 113:297–306, 2000.

[16] Kiehl TR, Mattheyses T, and Simmons M. Hybrid simulation of cellular behavior. *Bioinformatics*, 20:316–322, 2004.

[17] Munsky B and Khammash M. The finite state projection approach for the analyses of stochastic noise in gene networks. *IEEE Transactions on Automatic Control, and IEEE Transactions on Circuits and Systems I, Joint Special Issue on Systems Biology*, 53:201–214, 2008.

[18] Drewart B, Lawson M, Petzold L, and Khammash M. The diffusive finite state projection algorithm for efficient simulation of the stochastic reaction-diffusion master equation. *Journal of Chemical Physics*, 132(7):74–101, 2010.

[19] Van Kampen NG. The expansion of the master equation. In *Advances in Chemical Physics*, Vol. 34, I. Prigogine and S. A. Rice (Editors), Chichester: Wiley, 2007.

[20] Scott M, Hwa T, and Ingalls B. Deterministic characterization of stochastic genetic circuits. *PNAS*, 104(18):7402–7407, 2007.

[21] Van Kampen NG. Stochastic differential equations *Physics Reports*, 24(3):171–228, 1976.

[22] Elf J and Ehrenberg M. Fast evaluation of fluctuations in biochemical networks with the linear noise approximation. *Genome Research*, 13(11):2475–2484, 2003.

[23] Moler C and Van Loan C. Nineteen dubious ways to compute the exponential of a matrix. *SIAM Review*, 20:801–836, 1978.

[24] Kim J, Bates DG, and Postlethwaite I. Evaluation of stochastic effects on biomolecular networks using the generalised Nyquist stability criterion. *IEEE Transactions on Automatic Control*, 53(8):1937–1941, 2008.

[25] Lestas I, Vinnicombe G, and Paulsson J. Fundamental limits on the suppression of molecular fluctuations. *Nature*, 467:174-178, 2010.

[26] Laub MT and Loomis WF. A molecular network that produces spontaneous oscillations in excitable cells of *Dictyostelium*. *Molecular Biology of the Cell*, 9:3521–3532, 1998.

[27] Ma L and Iglesias PA. Quantifying robustness of biochemical network models. *BMC Bioinformatics*, 3:38, 2002.

[28] Maeda M, Lu S, Shaulsky G, Miyazaki Y, Kuwayama H, Tanaka Y, Kuspa A, and Loomis WF. Periodic signaling controlled by an oscillatory circuit that includes protein kinases ERK2 and PKA. *Science*, 304(5672):875–878, 2004.

[29] Kim J, Bates DG, Postlethwaite I, Ma L, and Iglesias P. Robustness analysis of biochemical networks models. *IET Systems Biology*, 153(3):96–104, 2006.

[30] Soll DR, Yarger J and Mirick M. Stationary phase and the cell cycle of *Dictyostelium discoideum* in liquid nutrient medium. *Journal of Cell Science*, 20(3):513–523, 1976.

[31] Kim J, Heslop-Harrison P, Postlethwaite I and Bates DG. Stochastic noise and synchronisation during Dictyostelium aggregation make cAMP oscillations robust. *PLoS Computational Biology*, 3(11):e218, 2007.

[32] CompBio Group, Institute for Systems Biology (2006). http://magnet.systemsbiology.net/software/Dizzy.

[33] MATLAB®: The Language of Technical Computing. Natick: The MathWorks, Inc., 2007.

[34] Nagano S. Diffusion-assisted aggregation and synchronization in *Dictyostelium* discoideum. *Physical Review Letters*, 80:4826–4829, 1998.

[35] To TL, Henson MA, Herzog ED, and Doyle FJ. A molecular model for intercellular synchronization in the mammalian circadian clock. *Biophysical Journal*, 10.1529/biophysj.106.094086, 2007.

[36] Lauzeral J, Halloy J, and Goldbeter A. Desynchronization of cells on the developmental path triggers the formation of spiral waves of cAMP during *Dictyostelium* aggregation. *PNAS*, 94:9153–9158, 2007.

[37] Wolkenhauer O, Ullah M, Kolch W, and Cho KH. Modeling and simulation of intracellular dynamics: Choosing an appropriate framework. *IEEE Transactions on Nanobioscience*, 3:200–207, 2004.

Index